Industrial Design 25th Annual

Design Review

Industrial Design 25th Annual

Design Review

Compiled and written by Edward K. Carpenter

Whitney Library of Design, an imprint of
Watson-Guptill Publications/New York

CONTENTS

Foreword by George T. Finley, 7

Introduction: A Tribute to Charles Eames, 9

Consumer Products, 45

Contract and Residential, 69

Equipment and Instrumentation, 81

Visual Communications, 111

Environment, 155

Credits, 183

Directory of Designers, 184

Directory of Products, 186

Index, 189

Foreword

This fourth edition of *Design Review* published in book form contains two significant changes from the earlier editions. One is the inclusion of prototype designs in the selections. The other is a major essay by Edward K. Carpenter on the work of Charles Eames, who died in 1978, the year covered by the review.

Both of these modifications were made in hopes of broadening the scope of the book and thus increasing its resource value. They were also motivated by our desire to commemorate in a meaningful way the 25th anniversary of Design Review (the first 21 having been published in *Industrial Design* magazine).

While the value of an essay on a designer the caliber of Eames is self-evident, some explanation is in order concerning the prototypes. Before deciding to include them in the review, we did an informal survey of designers to guide us. The responses were divided into two nearly equal groups. On the one hand were those who said the prototypes would make the review more exciting and would represent more of a leading edge of design. Some proponents said, for example, that the best solutions are often not produced because of a conservative client or one who simply does not appreciate "good" design.

On the other hand, opposition was based on a feeling that prototypes are simply not in the same category with produced designs and that it is unfair to place them side by side. Designs that have been manufactured or built must adhere to strict marketing, manufacturing, and cost restrictions which do not apply to prototypes. Finally, it was said that designers working for real clients don't have prototypes to submit because their best designs are, in fact, produced.

While the opposition is certainly not without merit, we decided that the potential value outweighed the risks, but that the prototypes should be clearly identified as such. The fact that only 4 out of the 81 total projects included are prototypes indicates the scrutiny which the judges applied in their evaluation.

When changes such as those we've outlined are made to an annual volume, the patience and skills of all those involved are given an added test. Special thanks for their contribution under these circumstances go to Sarah Bodine and Susan Davis of the Whitney Library of Design and to Susan Goodkind and Lorraine Robinson of *Industrial Design*. The jurors, as we have come to expect, performed beautifully.

George T. Finley
Editor
Industrial Design

INTRODUCTION

A TRIBUTE TO
CHARLES EAMES

In 1978 industrial design lost one of its preeminent practitioners, Charles Eames. Long a leader of the profession, recipient of five honorary degrees and innumerable major awards, designer of more than 70 films, a host of furniture, exhibits, toys, slide shows, and a molded plywood splint used by the Navy during World War II, Eames brought an exquisite sense of form and a perfectionist's dedication to his lifelong quest for structural clarity. His resulting designs have a purity that lets them escape what is called "their own style." That they are the product of Eames' hand and mind becomes apparent only in the clarity of their solutions and the obviousness of their quality. A look at the way Eames worked, how he thought of himself, and, briefly, what he designed is perhaps as instructive as a look at a cross section of 1978's industrial design, found later in this book. To understand Eames and his work is to understand other work better—to see it more clearly. That ultimately was what Eames wanted: to shape the way we see.

Charles Eames was 71 when he died last summer in St. Louis, the city of his birth. It is perhaps significant that Eames's roots were Midwestern, that he grew up in the heartland, studying and teaching there as a young man, taking with him when he went west in 1941 the heartland's values and attitudes. All his life he worked hard, reaping his pleasures from his work and plowing them back into it, living in a way that was as unostentatious and direct as the flow of the Midwest's prairies and rivers. "Can't pleasures be useful?" Eames would ask rhetorically. "I think that this is the objective: to get as many of the rewards of life from the work that you do."

"Charles took his pleasures seriously," recalls one of the people who worked with him. "Having a good time was a serious business." So was his business serious, something he worked at 12 hours a day, seven days a week. Often, especially when he was younger, the days turned into marathon sessions in which everyone in the office worked until, as a colleague recalls, "We were walking around like zombies." But despite physical and emotional exhaustion, Eames loved working this way. "We start gradually, as in a love affair," he once explained. "We only begin to ignite when every living moment is dedicated to the search."

John Neuhart recalls joining the Eames office as a young graphic designer and suddenly being confronted with a 32-hour work binge. "It took days to recover." As Eames got older, he found he could no longer work like that. After the 1964 New York World's Fair (for which he did a multiscreen film, some puppet shows, and the IBM pavilion with Eero Saarinen & Associates), he rarely stayed in the studio past 10 o'clock at night.

Eames' father had been a photographer and painter as well as a part-time private detective, who frequently worked undercover, writing young Charles letters signed "Victor Samuels." It may have been this exposure to sleuthing that gave Charles a partial hankering for anonymity. For years his shop has been in a converted garage in Venice, California, in a dingy building that still says "24-hour Towing Service" and "Fenders and Body Work" on the outside. He may have thought of this faded lettering as protective coloration, as a disguise. It was as if behind it he might find the solitude he needed for concentration on his work.

Eames worked only on projects that interested him, at his own pace. George Nelson, who collaborated with and commissioned Eames occasionally through the years, recently listed three of Eames' strengths as a designer. Number three on Nelson's list is that Eames "knew how to find clients, or manipulate clients, to let him do what he wanted to do."

Eames quoted the *Bhagavad Gita* to explain why he tried to work without client pressure. He'd gone to India in 1958 to advise the Indian government, offering recommendations that resulted in the creation of the National Design Institute at Ahmedabad, and facets of the East stayed with him. Eames' version of the Gita message is: "Work done with anxiety about the results is far inferior to work done in the calm of self surrender."

Inside, Eames' studio was more craft shop than industrial design office. Missing were the formal waiting room with plaques and photographs of the owner's designs. Missing too were offices, though Eames had an enclosed space which he rarely used. Instead he was always moving through the studio's large, brightly lighted loft spaces, among the 20 or 30 people who in later years worked with him there, conferring about graphic design, or furniture, modifying a model, adjusting lights or camera settings for a photography session. The studio lofts held power tools, molds, photography darkrooms, drafting boards, models and mock-ups, and the huge accumulation of toys, tops, trains, dolls, steamships, games, banners, and what *Fortune* once described to its corporate readers as "other quirkish oddities." Eames was an inveterate accumulator and collector. Once the airlines started using jets, he traveled widely, bringing back objects and slides from his assignments. Besides, every time he did an exhibit or a film, bits and pieces were left over, "a fantastic residue," an associate called it, which could be assimilated and reused.

Eames, of course, ran the office as a partnership with his wife Ray, who shared his passion for collection and his greater passion for form and structure. Throughout his career, while working on an amazing range of projects, Eames always thought of himself as an architect. His earliest study, training, and practice had been in architecture, and when people would ask him what he was, wanting to label his talent, he would say he was an architect, that he worked with structure. "I look at the problems around as problems of structure," he told them. He saw himself, too, as a problem solver, someone who like Marie and Pierre Curie, combing through a ton of pitchblende to isolate a tenth of a gram of radium, would take an awesome mass of information, sort and sift it until the relationships of its fragments became clear. "You try to find functional relationships," he said. "And you build a structure out of them. . . . Architecture should be able to concentrate and attack a puzzle."

Ray saw and supported this sorting process. She recognized it as one of

Charles and Ray Eames at work on their movie, Toccata for Toy Train, *1957. All photographs on pages 9 through 43, unless otherwise credited, are courtesy of the Office of Charles and Ray Eames.*

Charles's strengths. And she spoke of another: "His sense of form is to me perfect," she said. It is not surprising that Charles saw the same strengths in Ray. "She has," he once explained, "a very good sense of what gives an idea, or form, or piece of sculpture its character, of how its relationships are formed. She can see when there is a wrong mix of ideas or materials, where the division between two ideas isn't clear. If this sounds like a structural or architectural idea, it is. But it's come to Ray through her painting. Hans Hoffman was a great teacher. Hans worked with Ray for a long time, and any student of Hans's has a sense of this kind of structure."

Hollywood film director Billy Wilder became a friend of the Eameses. They had done a montage (of Lindbergh's plane being built) for Wilder's 1956 film, *The Spirit of St. Louis.* And though the Eameses partnership was so intense and intimate that it is hard to analyze, Wilder took a crack at it once: "Ray's imbued with. . . . absolutely perfect taste. She is also, I think, a very good organizer—she's much less of a dreamer than Charles is. I think she sort of holds things together." Physicist Philip Morrison, who worked with Eames on scientific films, said this: "Ray makes a surround around Charles, and Charles makes a direction for Ray."

But if Ray was the organizer, Charles handled the clients. "No matter how long you'd been there you never saw a client," recalls John Neuman. "All you had to do was your best possible work." Eames was a perfectionist. He demanded perfection from himself and expected it from his staff. "Once a person accommodated to Eames' view he was treated as a peer," says Neuman. "There was little of the master-pupil relationship. Charles would push you into areas you had never worked in before. 'If you demonstrated intelligence in one area you could in another,' was his conviction. He was impatient with the 'I don't know anything about art' approach often found in consultants."

Don Albinson, one of Eames' students at Cranbrook Academy of Art in 1940 and, after the war, an Eames' colleague for 13 years in California, says that Eames' "talent was in organizing—in leading, pushing, driving you to do what he wanted." Eames usually knew what he wanted and had the ability and enthusiasm to communi-

cate it. In fact, second on George Nelson's list of Eames' strengths as a designer is that he was "very good at stating what he thought he was doing."

Though Eames was good at describing a direction or depicting a goal, his delivery of that information could be disconcerting. "His speech would sometimes be slow and hesitant," says John Neuhart. "His sentences would trail off into nothing, making it difficult for a newcomer in the office or someone meeting him for the first time to know what he wanted." When this happened to Neuhart, he would go to Albinson, who'd been there longer, and Albinson would interpret: "He wants you to investigate this . . . or this. . . ."

While the Eameses' search was for relationships and structure, it was also for new forms, new materials, new ways of doing things. And it was a mark of their collective genius that the freshness of what they produced did not keep it from being accepted in the marketplace, from making money. As would anyone whose father died when he was twelve and who was brought up by a French mother, two aunts, and an older sister to be imbued with the idea that a *man* worked, Eames knew the value of money. He started working young, at an array of jobs, after school, summers, and when his furniture began to sell, he took royalties on all of it. *Fortune* estimated that by early 1975, 5.5 million pieces of Eames' furniture had sold, garnering Eames a 1.5 percent royalty on each piece. By 1974, the magazine guessed, Eames was receiving $15,000 a month in furniture royalties. The rest of the office gross came from fees and retainers. For a while IBM paid Eames a retainer to give its projects priority. And in the normal course of work Eames charged a client for payroll expenses plus sums for overhead and profit. During 1974, according to *Fortune*'s figuring, the Eames office took in $750,000, of which $400,000 came from IBM and the U.S. government for an exhibition on Franklin and Jefferson, the rest in fees and royalties. *Fortune*'s concern with money was academic, for the Eameses' way of life was subdued, their style a studied casualness, and in any given year, whatever money came in, they put back into the business, which was, of course, their life.

Eames' approach to cars is indicative both of his studied casualness and

of his love of objects. His cars were always black, always convertibles, not too ugly, not too ostentatious. So were Ray's. He liked to drive through the California sunshine on his daily trips from home in Pacific Palisades to the studio in Venice with the top down. He would drive a car to exhaustion, part with it reluctantly. Don Albinson remembers going with Eames to the Ford dealership in the mid-50s to turn in a car he'd bought shortly after World War II and pick up a new one. As they drove out of the showroom, Eames said ruefully, "It's not as nice as the old one." But he drove it for 18 years.

Like many designers the Eameses designed their entire environment, creating surroundings that nourished them, that sustained their image. They saw to it that each facet of their surroundings suited this environment exactly—car, clothes, studio, home. Just as the Venice studio was filled with collections and objects, so was their home—with plants, toys, photographs, shells, candelabras, books, driftwood, pillows—each possession positioned exactly to present the image of casualness. "There was always the banner of casualness," says John Neuhart, "but nothing was casual." Don Albinson tells of how the Eameses prepared for an informal evening with Billy Wilder: by having some of the office staff go to the house beforehand and arrange things so it would appear casual when Wilder came by. The staff set up the right color candles in the right places, making sure they had burned down to an appropriate length, and placed pillows in proper places. Everything was nicely thought out . . . to be casual.

Eames' clothes too had this studied casualness. His clothes sense was perhaps refined by living in a French household in the American Midwest, or perhaps, as Don Albinson notes, merely by good taste and judgment, "He didn't buy clothes to impress anybody; he was very careful not to be modish or arty about clothes." His dress was an informal mix of quiet colors and rough textures, like tweeds or corduroy. Eames would wear a vest, a tweed jacket, or a sweater draped over his shoulders, and carry a magnifying loupe around his neck, the way a film director carries a viewing scope or a doctor a stethoscope. A colorful Western bandana was always knotted at his throat or protruding from a pocket.

"Their clothes were like a uniform," recalls Don Albinson.

Ray always dresses in full skirts with capacious pockets. Like Mrs. Glass, matriarch in J.D. Salinger's stories, Ray can produce from her pockets the right thing at the right time, anything from a power tool to a match.

But perhaps the most important aspect of Eames' clothes was that in texture, color, and cut they had a kinship with his work. Aside from his suits, which he had tailored in London to last, his clothes often had a crafted look, as if they came off a loom in Mexico or were sewn by a seamstress friend. This handcrafted look appears in his work: in his exhibits, toys, furniture, even his films. This look, in carefully machined objects, distinguishes his designs, giving them a simplicity that goes back to early America, to things made with tools, not machines. Eames loved tools. John Neuhart remembers Eames giving him a little lecture about how to sharpen a pencil with a pen knife. Eames sharpened his pencils with a knife. He wouldn't use an electric pencil sharpener, even a hand-cranked one.

While Eames' designs are ultimately turned out by machines, their appearances do not evoke the machine or pay homage to it. His work has a craft simplicity more than a machine simplicity.

Once when Eames was put on the spot by some students at Oxford University, who asked if he thought of himself as an artist, he said he thought of himself as a tradesman. He did not, he told them, think that was incompatible with being an architect, that architects are tradesmen. The word "architect" "implies a structure, a kind of analysis as well as a kind of tradition behind it. . . . The tools we use are often connected with the arts, but we use them to solve very specific problems."

"Any good tradesman," he went on, "should work only on problems that come within his genuine interest, and you solve a problem for your client where your two interests overlap. You would do an injustice to both yourself and your client to work on a problem of interest to the client but not to you— it just wouldn't work. In that spirit we are tradesmen. If your work is good enough, it can be art, but art isn't a product, it's a quality. Sometimes that's lost sight of. Quality can be in anything."

First on George Nelson's list of Eames' strengths as a designer is that he was an "extraordinary craftsman."

In a way the house Charles built for himself and Ray, on a bluff, backed up against a hillside, overlooking the Pacific in Santa Monica's Pacific Palisades, was crafted. (The house was one of two case study houses sponsored by the West Coast magazine *Arts and Architecture* in 1949. The other one was built for the magazine's then editor, John Entenza, sited farther down the bluff, below the Eames house and designed by Eames and Eero Saarinen.) Its parts were machined, and that was the whole point of the house, that it be made of readily available, machined, off-the-shelf parts, components that anyone could assemble and put together. But in the way these parts are put together Eames' house looks crafted. The way Eames juxtaposed stucco, glass, and translucent panels, without repeating uniform elements, using industrial sash, wires, clamps, bolts, and trusses, has nothing to do with machines. Just as important to that crafted effect—to the environment that the house became—was what Eames and Ray filled it with. "Functioning decoration" Eames called the lovingly positioned collections of shells, baskets, blankets, mats, candles, pillows, pots, plants, sculpture, chairs. These interior objects add texture, form space, play with the light that comes subtly, with constant change, through the house's transparent and translucent shell—a light further controlled by draperies. Five years after they moved into the house Eames made a film of it. A succession of quick, still photos, *The House* shows clearly that interior collection and arrangement, as much as the architecture, make the house what it is. The film's images are those you would see if you stood in the house and looked about it. *Architectural Design*, the British magazine, called them "aspects which make up the inhabited place: window, floor, bowl of fruit, staircase, flowers, dishes on the table, eucalyptus trees." The house is a happy blend of structure, space, and content.

Eames went out of his way to make the house look light, to make the walls, floor, and roof thin, to have each element appear to rest gently against another, with no seeming support. Though the house is more like a house of cards, which it has been called, than

a kite, which it has also been compared with, it is unquestionably substantial, its edges, corners, supports, and foundation all firm, rigid despite their appearance.

By using wires (unobtrusively placed) for cross bracing, Eames avoided a heavy entablature. His roof is composed of paper-thin steel trough decking, topped by equally thin insulation board and supported by light 12" (30.5 cm) deep open web trusses. These trusses are bolted to the 4 x 4" (10.2 x 10.2 cm) steel sections which form the wall's framework. Window systems bolt into this framework and so do sections of 1½" (3.8 cm) steel siding covered with stucco. This structural arrangement lightens the interior space and enables windows to begin at the roofline's outer edge, so the effect of lightness approaching delicacy is apparent inside and out. That the house is a box is apparent. It is obviously a box in shape and in concept. Eames meant it as a container that would set off and enhance the things contained. But he kept it from appearing boxy by making it two boxes. One is for living quarters, the other for a workshop. The two were originally connected by a courtyard, in later years by this courtyard and a board walkway.

Actually the house that Eames built was a second design. The first one, done in 1947, shows a house raised on steel stilts one story above the ground, meeting the bluff in back of it at right angles, This design didn't satisfy Eames. Materials for the house—the frames, supports, and fastenings—sat in the yard for a year while he "let the design float." This period of mental fermentation was typical of Eames, and so was the way the house looked in his first design. "Let it float," he would say to his associates, meaning don't lock in a design too soon, keep working with it until it jells, until it's ready to come forth full blown. "Only Charles could tell when a design was ready," remembers John Neuhart. But the idea of letting a design float is just as applicable to his designs' appearance as to their conception. Many of his chairs seem to float in space, their legs minimal and unobtrusive. And the same is true of his home's first design. Except for its connection at one end to the bank behind, the house would have seemed to float unattached to the ground, gripped like his chairs by the surrounding air.

Eames House, Santa Monica, California, 1949. "Its parts were machined, and that was the whole point of the house, that it be made of readily available, machined, off-the-shelf parts, components that anyone could assemble and put together. But in the way these parts are put together Eames' house looks crafted. The way Eames juxtaposed stucco, glass, and translucent panels, without repeating uniform elements, using industrial sash, wires, clamps, bolts, and trusses, has nothing to do with machines. Just as important to that crafted effect—to the environment that the house became—was what Eames and Ray filled it with. 'Functioning decoration,' Eames called the lovingly positioned collections of shells, baskets, blankets, mats, candles, pillows, pots, plants, sculpture, chairs. These interior objects add texture, form space, play with the light that comes subtly, with constant change, through the house's transparent and translucent shell—a light further controlled by draperies."

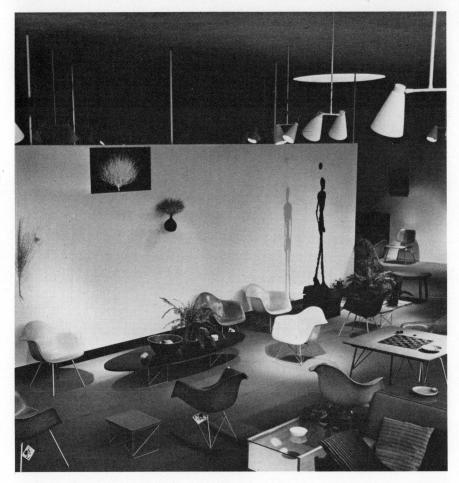

Herman Miller Company showroom, Los Angeles, 1949.

Though the precedents for the Eameses' house are Japanese—its near delicacy, the way it focuses light and shadow, bringing indoors and outdoors together harmoniously—it can be argued that its roots are also European. In its tracery of steel and glass are echoes of houses built in California by Neutra and Shindler who arrived from Europe in the 30s and of designs by Raphael Soriano, whose simple, steel-framed houses drew on Neutra and Shindler. (The Eameses lived in a Neutra-designed apartment building on arriving in California in 1941.) And critics have seen precedent too for building with standardized industrial components. Bernard Maybeck fabricated a good deal of the exterior of his First Church of Christ, Scientist, in Berkeley from galvanized sash and asbestos.

Perhaps more than antecedents, California's stylistic freedom, its lack of cultural constraints, shaped the house. Eames said in the 50s, "Los Angeles is receptive to new ideas, both good and bad, because it has as yet no cultural standards into which things must fit."

In 1927 Eames had flunked out of Washington University's architectural program; so the delight his house gave him, the attention its design received in the architectural press, must have seemed vindication for the university's shortsightedness. His troubles at the university came from inclinations that with maturity were to help bring him fame and satisfaction—inclinations for hard work and fresh, natural design.

Even in 1927 Eames' schedule was awesome. He found it difficult to combine work in the St. Louis architectural firm Trueblood & Graff with classroom studies, especially when those studies were Beaux Arts and Eames' hero then was Wright. The formal studies ceased temporarily, but he found ways to continue studying on his own: as a designer for Trueblood & Graff; during a 1929 visit to Europe; in his own firm Gray and Eames, which he opened in St. Louis in 1930; in Mexico where he lived for a year; then in another firm, Eames & Walsh, in St. Louis in 1935.

But his own house and the Los Angeles showroom he designed for the Herman Miller Company about the same time were his last ventures into formal architecture. Similar to the house, in that it had a street wall of clear glass and opaque panels mounted in industrial steel frames, the Herman Miller showroom had carefully arranged interiors of Herman Miller furniture and Oriental objects. Though Eames designed a house for Billy Wilder and no doubt would have seen it built had not Wilder taken his pleasure entirely from the model, Eames' concentration went into other things.

During the 40s that concentration had never been far from chairs, though he designed other pieces of furniture: tables, storage units, a marvelously undulating folding plywood screen, and, during the war, the molded plywood splint. He saw chairs as architecture, of course. Chairs were merely a more concentrated problem than a house. "The chairs are architecture," he once said. "They have structure just as the front page of a newspaper has structure. The chairs are literally like architecture in miniature. For an architect who has difficulty controlling a building because of the contractor and the various forces brought to bear on anything that costs that much money, a chair is almost handleable on a human scale, and so you find great architects turning to chairs: Frank Lloyd Wright, Mies van der Rohe, Le Corbusier, Aalto, Eero Saarinen—any number of them doing it, because this is architecture you can get your hands on."

There was more to it than that. Some architects wanted to control the interior environments their architecture only partially created, and they designed much of their furniture for special situations within their buildings. It is hard to envision, for instance, Mies's Barcelona chair in an average American living room. Eames' chairs, not fettered by this restriction, can stand well on their own or fit equally well, like worldly visitors, into almost any situation. It is as easy to imagine an Eames side chair enhancing a townhouse as it is to picture it at home in a trailer; this design versatility is behind the chair's success.

In designing his chairs Eames worked with his hands, or at least with tools, which are extensions of hands. A chair's design started with a verbal concept. For 13 years after the war his associate most involved in furniture was Don Albinson, who had worked with Eames on furniture as his student in the 30s. (Harry Bertoia, too, was in the Eames shop for a while working on furniture; so, briefly, were others.) One of Eames' great qualities, Albin-

son maintains, was "always knowing where he wanted a design to go." But his instructions were oral: he never used paper or pencil to point a direction. Eames would talk about the piece of furniture he had in mind "sketching it verbally," says Albinson, and Albinson, or Bertoia, would go off and build a model. These models were, as far as possible, constructed from the materials envisioned for the final design. Then the Eameses and their associates would refine the model, carving a piece off a leg, changing a seat angle, until Eames considered the design ready. They would sit in the models, testing them for comfort, which may be why Eames chairs are as easy to sit in as they are to look at. Eames would work through a seemingly endless progression of models, refining, working on certain aspects, such as arms. Albinson recalls doing 13 different arms for the lounge chair, none of which pleased Eames. Finally with Herman Miller complaining about time and expense, Eames reached back and picked an arm they had tried three and a half months earlier.

This perfectionism often made Eames exasperating to work for. Albinson recalls that Eames never seemed satisfied with a final design, always looking as if the design failed to match an inner vision of perfection. Like Jerry Kramer, a professional football player under Vince Lombardi, who, after being yelled at by Lombardi for a season, was so convinced he could do nothing right that disbelief was his only reaction to being named all-pro, Albinson was always left with a feeling he'd failed, that Eames wasn't really satisfied with what he'd done. "Thirteen years as a failure," he says today with a little laugh.

Accepting a fellowship to study at the Cranbrook Academy of Art in 1937, Eames went to Michigan and stayed to teach, as head of Cranbrook's department of experimental design. In 1940 he and Eero Saarinen, whose father Eliel then ran Cranbrook, entered a furniture competition at New York's Museum of Modern Art. Eliot Noyes, director of the museum's department of industrial design, had set up the competition for "Organic Design in Home Furnishings," and Eames and Saarinen took first prize for seating and other living room furniture.

"One of the things we had committed ourselves to," Eames was to say in the 1970s about the chairs he de-

signed with Saarinen in 1940, "was trying to do a chair with a hard surface that was as comfortable as it could be in relation to the human body and also that would be self-explanatory as you looked at it—no mysteries—so that the techniques of how it was made would be part of the aesthetics. We felt very strongly about this, because at the time there were so many things made with the opposite idea in mind, that is, to disguise a thing as if it were made at the Gobelin factories in Paris, when in fact it had been manufactured by modern techniques."

Though everyone connected with the competition was eager to see this furniture into production and though Eames and Don Albinson moved for a while into the plant of the Haskelit Company, which had been selected to mold the plywood chair shells, they never found a successful way to mechanize production.

Despite Haskelit's inability to produce the chairs, Eames characteristically refused to release the problem. In 1941 he married Ray Kaiser, who had helped Eames and Saarinen prepare their entries for the museum competition, and moved to California. In a room of their rented apartment they continued experimenting with plywood molding. Eames worked by day as a set designer for MGM, returning at night to search for ways to mold plywood in two planes. Out of their experiments came the molded plywood splint used by the Navy during the war, but chairs were always on Eames' mind. By the time Don Albinson got out of the service in 1946 and went to work with Eames, Eames had abandoned the idea of molding his chairs in one piece, deciding instead to mold seat and back as separate shells, or "petals," as he called them. He experimented, too, with steel rods for legs, giving the chairs a lighter look. The problem became one of attaching the metal to the wood. They experimented using a rubber disc as a joint between metal legs and wooden shells, and Albinson remembers, through the haze of 30 years, a mountain of a man named Norman Bruns, who thought he had the answer. As Albinson tells it, Bruns looked like a cross between an unsuccessful prize fighter and the current comic book hero, The Incredible Hulk. His speech was halting and husky. "I think we can bond this with the electronic bonding technique," he would say, go off and return with what Albinson describes

A3501

SIDE CHAIR

PLYWOOD

ALUMINUM

PLYWOOD
RUBBER WELD

ALUMINUM

FULL SIZE CONSTRUCTION DETAIL
OF LEG CONNECTION

28"

ONE QUARTER FULL SIZE

"Organic Design in Home Furnishings," Charles Eames and Eero Saarinen competition drawings, Museum of Modern Art furniture competition, 1940. Photographs courtesy Museum of Modern Art.

as "$5,000 worth of equipment which gave off sparks and zapping sounds and turned the plywood to charcoal." They ended up gluing the rubber discs to the wood and sealing the bond with heat.

In 1946 the Museum of Modern Art exhibited Eames' new chairs. Eames thought that would be the end of it, that there would be no way to get them produced, and indeed it almost was the end. Eames and Ray were thinking of seeking employment as circus clowns, goes one version of the story, when George Nelson convinced the Herman Miller Company they needed Eames chairs. The company adopted Eames, despite Mr. Herman Miller's reported initial reaction to Eames' designs—"That's never going to go into any showroom that has my name on it."

Furniture ideas often came to Eames from the suggestions of his friends. His aluminum indoor-outdoor furniture group, for instance, was inspired by a comment of Alexander Girard. In furnishing the Columbus, Indiana, house that Eero Saarinen designed for Irwin Miller, Girard noted that there was no really good outdoor furniture. And the elegant aluminum chaise that appeared in 1968 with its padded leather cushions evolved from a need felt by Billy Wilder for an office couch on which he could nap after lunch. Some 6' (1.8 m) long and 28" (71 cm) off the floor, the chaise, which has no arm rests, is only 16" (40.6 cm) wide. Wilder could only lie on it with his arms folded over his chest, and after twenty minutes, his arms, slipping, would wake him to the afternoon's work.

By 1959, thirteen years after Herman Miller first turned out an Eames chair, when Albinson finally left the Eames office, Eames' interest in furniture was dimming. "It got so in the end I had trouble getting his decisions on designs," recalls Albinson. "He'd be involved with films or exhibits, and I'd have to catch him at dinner; even then he often wouldn't remember what we'd discussed from one meeting to the next."

A case can be made that after 1959, when Albinson left, Eames rarely again did any original furniture design, that almost everything, with the possible exceptions of his tandem airport seating and his steel and leather chaise, were refinements of ideas he had had years before. Even the chaise may have stemmed from an earlier Eames concept. Arthur Drexler, director of the Museum of Modern Art's department of architecture and design, who put on an Eames furniture show in 1973, sees antecedents for the chaise in an experimental molded plywood and steel chaise Eames worked on in the 40s. Well before 1959 Eames was reaching into his past for furniture design ideas. The lounge chair developed in 1956 as a TV chair for Billy Wilder, probably one of Eames' most famous designs, borrowed its three molded plywood sections from an experimental lounge chair of the 40s. Albinson remembers returning from the service to work on this early chair. The "Donaster," Eames called it jokingly, never completely satisfied with it, and indeed he spoke with the same loving dissatisfaction of the refinement: "The black leather downfilled lounge chair, I have never considered as good a solution as the other one (the steel rod and plywood petal dining chair), although it has apparently given a lot of pleasure to people. It has a sort of ugliness to it. I take a certain comfort in the fact that Picasso pointed out through Gertrude Stein (in the Alice B. Toklas book) that anything that is truly original must have a touch of ugliness because it has not had a chance to be honed to the elegance that comes with refinement."

Perhaps by drawing upon past ideas Eames was putting his furniture through a process of refinement that came from letting them "float" for years. At its best his furniture has a distinctive lightness. It appears as if it were floating above the floor or sometimes as if it were held off it, like a piece of sculpture on a pedestal, for inspection and appreciation. At the same time it is inviting; it looks comfortable; it makes you want to sit down and stay put.

There seemed to be a direct conveyor running from the Eamses studio in Venice to the Museum of Modern Art in New York. No sooner had the Eames office designed a piece of furniture than it would show up in the museum collection. Today the museum holds some 50 different pieces of Eames furniture. In his introduction to the 1973 Eames furniture show, Arthur Drexler called Eames the most original American furniture designer since Duncan Phyfe. Phyfe died in 1854.

A3501

CONVERSATION

Opposite page: Plywood chair, metal legs, 1946.
Left: Plastic arm chair, 1949.
Below: Arm chair on cross-rod base, drawing for Museum of Modern Art "Good Design Show," 1950.

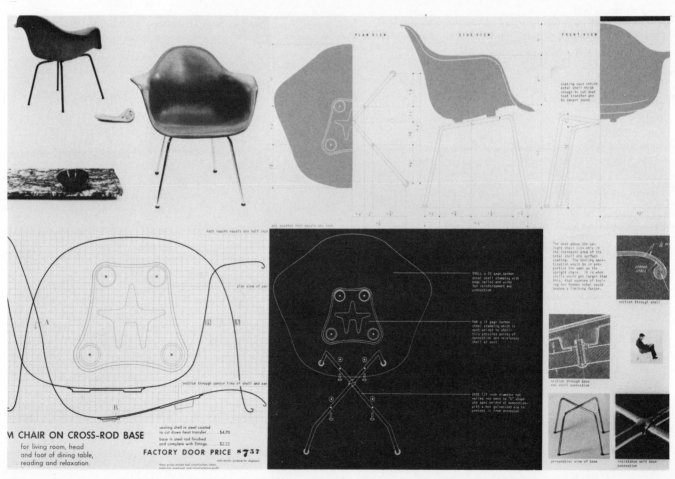

Right: Detail, La Fonda chair, 1960.
Below: Upholstered plastic chair, 1950.

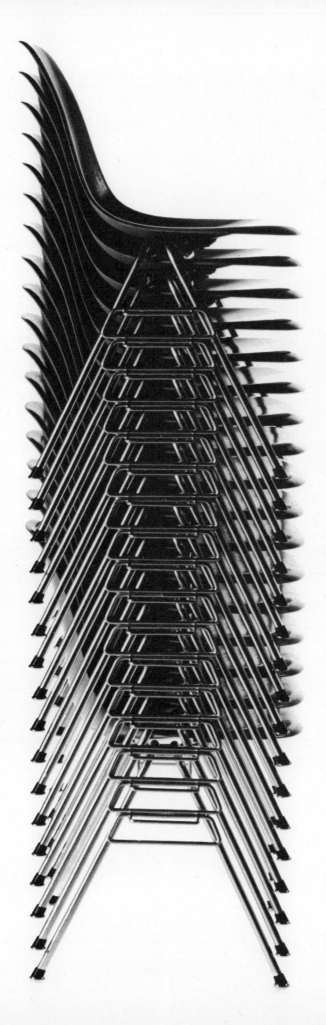

Stacking chairs, 1954.

Lounge chair and ottoman, 1956, on permanent exhibition at the Museum of Modern Art. Photograph courtesy Carl Ruff Associates.

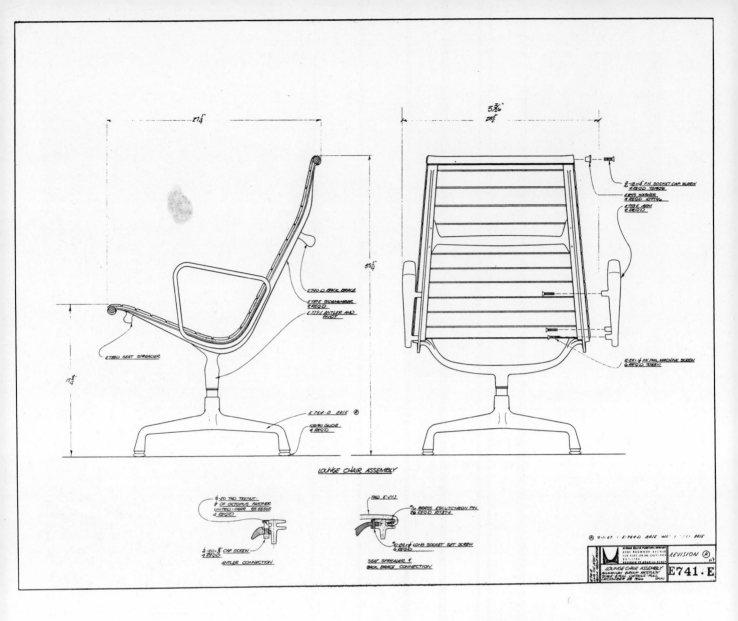

LOUNGE CHAIR ASSEMBLY

ANTLER CONNECTION

SEAT SPREADER &
BACK BRACE CONNECTION

E741·E

Above: Lounge chair assembly, 1967.
Right: Detail, tandem seating, Dulles Airport, 1962.

Soft pad group, 1969.

Chaise, 1968.

By 1959 Eames had moved heavily into films. He took them on as a problem in design, or as he would have said, architecture, for essentially the same reason he had set aside buildings for furniture. It was a matter of control. Even more than in the design of a chair Eames could control the look and structure of a film. And perhaps most important, with film he could control the statement, both the idea and the way it was presented.

"Eames turned to movies because he could most fully control the statement he was making. He wanted to control what he said and did completely. He wanted to avoid the embarrassment or hurt that came to himself or his clients when his work was misconstrued. He saw this as part of his responsibility," says Don Albinson. And Eames himself has said , "You can make statements on film that you just can't make any other way, certainly not in designing a product, not even in writing a book. You have certain elements of control—over the image, the content, the timing— that you can't have in other media."

Eames recognized, of course, that even with film that control was not complete; it was merely more complete than in other media. "We have fallen for the illusion that film is a perfectly controlled medium; that after the mess of production, when it is all in the can, nothing can erode it—the image, the color, the timing, the sound, everything is under control. It is just an illusion. Thoughtless reproduction, projection, and presentation turn it into a mess again. Still, putting an idea on film provides the ideal discipline for whittling that idea down to size."

The Eameses first made a film, they said, because a friend left them in charge of a 16mm projector, and they had nothing to show on it. Their initial movie, *Traveling Boy*, done in 1950, is a journey through the world of toys, guided by a mechanical boy. Eames loved toys and always had lots of them around; many he designed himself. "Toys are less self-conscious than any other design form," he said. "Toys are really not for children; they're for adults, especially grandparents." Eames focused on toys frequently in his first years of filmmaking. And in a way, as Paul Scharder pointed out in a review of Eames' films in *Film Quar-*

terly, there was a connection between Eames' work in furniture and his films. "Through precise, visual, non-narrative examination the toy films reveal the definitive characteristics of commonplace objects. The toy films were the natural place for the Eameses to begin in film, for they found in simple, photographed objects—soapwater running over blacktop, toy towns and soldiers, bread—the characteristics they were trying to bring out in the furniture designs." The narration for *Toccata for Toy Trains,* a 14-minute film the Eameses did in 1957, says in part: "In a good old toy there is apt to be nothing self-conscious about the use of materials. What is wood is wood; what is tin is tin; and what is cast is beautifully cast."

Not long after the Eameses started experimenting with film, they got a call from Lamar Dodd, head of the department of fine arts at the University of Georgia. Early in the summer of 1952 Dodd had invited George Nelson to Athens, Georgia, to consult with the faculty on educational policy. Nelson didn't really want to go. It was hot enough in New York, and he explained to Dodd that he didn't know anything about educational policy. "I know," Dodd said. "That's why we want you."

After an initial consultation Nelson suggested they call in Eames to help. Eames, of course, didn't know anything about educational policy either. It was the kind of situation he, too, liked. Looking at the way fine arts were taught in Georgia, "it was our feeling that the most important thing to communicate to undergraduates was an awareness of relationships," Nelson wrote a year or so later. "Education like the thinking of the man in the street was sealed off into many compartments. If a girl wanted to know something about decorating her future home and what she got was a class in painting, this might make perfectly good sense, but perhaps it was up to the school to build a bridge between the two so that she might see how they were related." Nelson and Eames decided to put together a sample lesson for an imaginary course to illustrate their point and to call in Alexander Girard to help them do it. The subject of the lesson was to be grand. Communication, they called it. Girard agreed to make corresponding exhibits. Nelson and Eames split the one-hour lesson, each taking half, and the three set off, Nelson to New York,

Eames to California, and Girard to Michigan, to work on it.

Recently Nelson commented on how they worked on the project. "I recall a conversation on a train where we blocked it out. Fortunately when we fit the pieces together it worked."

There were lots of pieces. The idea was to flash seemingly unrelated phenomena before the students in a way that exposed relationships, and the achievement of this exposure, Eames and Nelson felt, called for films, slides, sounds, music, and narration. Five months later the three met in Athens with enough equipment, as Nelson put it, for a "traveling medicine show." They had slide projectors, screens, cans of film, boxes of slides, reels of magnetic tape, and Girard's exhibits, standing about in gigantic cases waiting to be unpacked. Girard had also brought bottles of scents, which he would later, like a pleasantly mad scientist, let loose in the air conditioning system at critical points in the show. Eames thought the odors quite effective: "They did two things: they came on cue and they heightened the illusion. It was quite interesting because in some scenes that didn't have smell cues, but only smell suggestions in the script, a few people felt they had smelled things—oil in the machinery, for example."

Odors were an important part of their presentation and so were sounds. "We used a lot of sound, sometimes carried to a high volume so you would actually feel the vibrations," Eames recalled years later. "So in a sense we were introducing sounds, smells, and a different kind of imagery. We did it because we wanted to heighten awareness."

It was multimedia. They were refining, if not inventing, multimedia events. Multimedia, of course, had been around for some time. "Any motion picture would be multimedia," Eames pointed out, "because it has both sound and images."

Their real innovation was multiscreen projection, an idea that occurred to them as they ran through slide sequences. Two slides shown at once might illustrate certain contrasts, they reasoned. From two it was an easy step to three screens and three simultaneous slides, producing, said Nelson, "sort of a poor man's Cinerama."

An inveterate photographer, Eames delighted in multiscreen projections. His father, of course, had been a photographer, as well as an artist and private eye, and shortly after his father's death in 1919 Eames came across photography equipment in the attic and started experimenting with wet plate photography. He kept taking pictures the rest of his life. Every time he returned to California after a trip he had hundreds of slides. Organized, these would find their way into projects or become references. "Films," Eames explained, "are an outlet for a picture-taking maniac." For a photography maniac multiple screen projection was like a nugget to a gold seeker. Eames used another image: "After you have made one multiple-image show, you're like a little boy who has been given a hammer—everything he encounters needs hammering."

Out of the "Rough Sketch for a Sample Lesson for a Hypothetical Course" came the *Communications Primer*, a film the Eameses made for IBM, which gave them a new direction. A *Communications Primer* sought to give architects insight into detecting and handling the masses of information they need to design cities for changing social situations and increased population. *Communications Primer* was the Eameses' first idea film, starting a string that treated science (*The Powers of Ten, The Lick Observatory, House of Science*) and technology (*Babbage, Schuetz Machine, Computer Glossary,* and so on).

But the first real chance they had to use multiple screens on a large scale came at the American National Exhibit in Moscow in 1959. Again George Nelson provided the chance. Nelson and his staff not only designed most of the exhibit but also installed it. Leading off was to be some sort of introductory statement about America and the fair, housed in a giant Buckminster Fuller dome, and this statement Nelson turned over to Eames, "because he was the only one I could think of." What emerged was a 12-minute presentation made up of films of 2,200 slides, shown on seven screens. The Eameses, of course, went off to California to make their film, and as Eliot Noyes once told it, the night before the exhibit was to open there was no sign of the Eameses, no one knew where they were or when they were coming. As a group stood, waiting in the vast emptiness of the Bucky Fuller dome, the seven screens and seven projectors in place, the Eameses materialized out of the gloom of the doorway, two far-off, tiny figures carrying shopping bags holding seven reels of film. "You'd think there would have to be some adjustments, some last minute changes," said Noyes. But the Eamese just fed their reels into the projectors and everything worked perfectly." Their film showed Americans going through what Walker Percy calls the "everydayness" of their lives, working, eating, going to church, playing with the kids. And it showed, too, the artifacts surrounding those lives, superhighways, cars, housing developments, TV sets, blenders. Eames wanted the film to end with a giant jet airliner taking off over the camera, roaring up into the sky, symbolizing something like progress or the way air travel has made the world smaller. But Ray said no. She wanted the camera instead to move in close, to focus down onto a bowl of flowers, of Forget-Me-Nots. And that, of course, is what they did. "Forget-Me-Nots," said Eames, "turned out to be called the same thing in every language, and the Russians would file out of the dome after the film, muttering 'forget-me-not, forget me not.'"

The way he took the work off to California, where he puzzled over it with his staff, was typical of Eames. He mulled over the problem in the solitude of his Venice workshop, far from the distractions of concerned kibitzing. Eames was relived, if a little surprised, that he'd been left so marvelously alone. He told Owen Gingerich, a Harvard astronomy professor who for many years was a consultant on Eames exhibits: "Theoretically it was a statement made by our State Department, and yet we did it entirely here and it was never seen by anyone from our government until they saw it in Moscow. It was a little touchy, but one of those things. If you ask for criticism, you get it. If you don't, there is a chance everyone will be too busy to worry about it."

Seemingly arbitrary, the use of seven screens was calculated. "We had the very difficult problem of making the first statement from this country to the Soviet Union since the Russian Revolution," said Eames much later. "We knew that words had their limitations; they could be used to the point of being almost without effect. . . . We knew that some images were well known to the Russians. If, for example, we were to show a freeway interchange, somebody would look at it and say, 'We have one at Smolensk

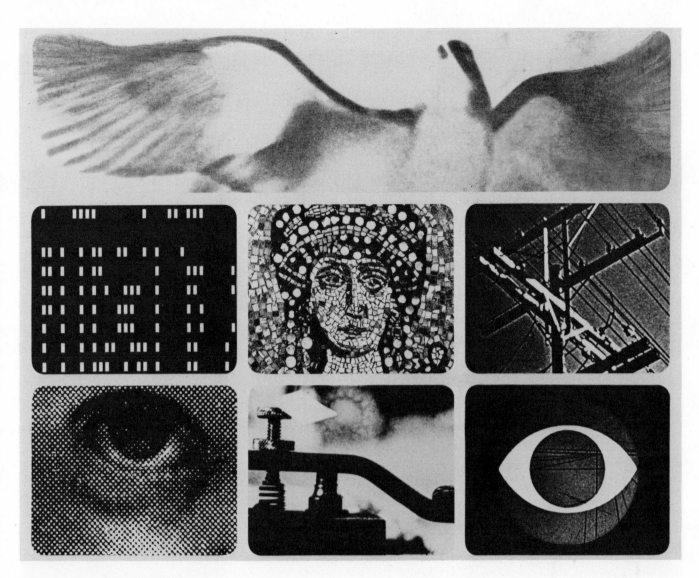

Communications Primer, *1953.*

Above: American National Exhibit, Moscow World's Fair, 1959.
Left: House of Science, 1962.

and one at Minsk; we have two, they have one.' So we conceived the idea of having the imagery come in multiple forms as in the 'Rough Sketch for a Sample Lesson.' "

But if not two, how many? Why seven? "We wanted to have a creditable number of images, but not so many that they couldn't be scanned in the time allotted. At the same time the number of images had to be large enough so people wouldn't be sure how many they'd seen. With four images, you always knew there were four, but by the time you got up to eight images you weren't quite sure. . . . They were very big images—the width across four of them was half the length of a football field."

Eames saw his commitment as one to information, not just effect, and he was disturbed by multimedia shows that evolved during the psychedelic years, shows that cut arbitrarily from image to image, offering only a pretense of information. "In planning the Moscow show, we tried out various tricks and rhythms in changing the images. We discovered that if you had seven images and changed one of them," said Eames, "this put an enormously wasteful, noninformative burden on the brain, because with every change the eye had to check every image to see which one had changed. When you're busy checking you don't absorb information. Franticness of cutting tends to degenerate the information quality."

Eames' multiscreen films culminated with his 22-screen presentation in the egg-shaped theater he and Eero Saarinen designed for IBM at the 1964 New York World's Fair. The film and its presentation were unquestionably exciting. But typically, Eames was trying to get across an idea, an idea this time about problem solving. In a way, the film was a 22-screen, color-and-sound statement of how Eames worked, of how he picked a problem apart to discover its facets and their relationships. Visitors to the IBM pavilion seated themselves at ground level in bleacher-like seats and the entire bleacher section was lifted hydraulically into an egg-shaped theater overhead. Then a host stepped out beneath the 22 screens.

"Ladies and gentlemen, welcome to the IBM information machine . . . a machine designed to help give you a lot of information in a very short time.

"It can help us look at things more

closely than we normally do.

"It can pick out and give special emphasis to things we often take for granted.

"Or it can offer a broad perspective of morning in Manhattan.

"The machine brings you information in much the same way as your mind gets it. In fragments and glimpses—sometimes relating to the same idea or incident.

"Like making a roast in the morning.

"Here is another example. Each screen shows one aspect of the notion—*road race!*"

Abruptly the roar of high-performance engines engulfs you; images on the screens bounce and tremble, as if you're riding behind the wheel of a car negotiating turns at 100 mph. Screens show fragments of the road; cars flash from one screen to the next.

Just as suddenly the screens go dark and the host goes on:

"That's how the information machine works. Now this is how we would like to use it. We'll show you some examples of problems that are very complex and some that are familiar to us all.

"You'll see that the *method* used today is solving even the most complicated problems is essentially the same method we all use daily.

"It is the method of attack that forms a link between our most sophisticated questions and our familiar ones. The recognition of this link can make us feel at home in this changing and complex world."

Eames started work on a film by sketching his thoughts. Unlike his furniture designs, which started with words, his films started with pictures. He would draw out a visual score for a film. John Neuhart remembers the score for a *Communications Primer* being drawn on a roll of adding machine paper. For years Glen Fleck elaborated on these drawings, preparing storyboards to guide the actual shooting, all of which Eames did himself.

Fleck, said Eames, "is one of the very few people who has a sense of what it is to communicate meaning. What is more, he has a sense of when he has not communicated it, and a sense of when he has not understood it in the first place—very rare."

Often Eames' films were beautiful, exquisite without being delicate or self-consciously pretty. Their beauty grew out of Eames' sense of form and composition and his technical mastery of the camera. Another Eames consultant, MIT physicist Philip Morrison, saw something further: "It seems to me that Eames' films celebrate the world. There were painters who did

that in medieval times and the early Renaissance, but no longer. There is this quality of love in the Eames films for the way the world is, the texture of wood grain, the crisp breaking of bread."

Eames films were thrilling and beautiful not only to the masses of people who saw them in Moscow, Seattle, and New York, but to the critics, too. His films won awards at the American, Edinburgh, Melbourne, San Francisco, Mannheim, Montreal, and London film festivals, and in 1960 his fast-cutting sequence of famous people who had died during the 50s, for the CBS special "The Fabulous Fifties," won an Emmy.

If the Eameses' techniques—of presenting a viewer with an abundance of information—worked in films, it was not entirely successful in exhibits. Here Eames' passion for collection may have overloaded a viewer's circuits, no matter how well the Eameses organized the material nor how attractively they displayed it. They arranged layer upon layer of material, drawing on their collections, on the bits and pieces of things in the workshop, avoiding clutter only by the rigor of organization. Ray Redheffer, the UCLA mathematician who worked with Eames on the brilliant exhibit "Mathematica" at the Los An-

geles Museum of Science and Industry in 1961, said that Eames "knows the right thing to have sitting around."

In exhibits it is hard to control the path a viewer takes through a designer's organization. Eames found that often when someone saw one of his films twice, the second time through he would see things he hadn't at first. He got so used to being told that he'd changed his films, that he'd added something since someone had seen it first, that he stopped denying it. "We hadn't added anything," he explained. "They'd merely taken a different route through it."

In an exhibit Eames had even less control of this route. Besides, a visitor could excape from an exhibit. For a film Eames had captive viewers, in the dark, with nothing else to watch. "Not even a senator dares to stand up and interrupt a film," he once noted. A viewer can't stop the film, skip ahead, turn it back, go on to something else, leave. All these options are open to an exhibit-goer. In a good exhibit he is bombarded with stimulation, he can see or hear what is coming next or what he has just passed and, besides, his path through an exhibit is often not rigorously guided; he can start at the end, if he wants, or in the middle. When an exhibit such as "Franklin and Jefferson," which the Eameses

IBM Exhibit, New York World's Fair, 1964. Theater design in collaboration with Eero Saarinen.

prepared with IBM funding as a Bicentennial exhibit, presented an abundance of material, demanding that a viewer read, for understanding, a forest of captions hanging on banner-like plaques from the ceiling, viewers' circuits overloaded. Like overheated calculators, they could not compute. They backed off and complained. "Franklin and Jefferson" became a book, though Eames thought of it more as a newspaper, as "a colored, walk-through tabloid," albeit a beautifully illustrated one with artifacts (a vacuum pump Franklin used in experiments, a stuffed buffalo, the Louisiana Purchase treaty), art, and photographs; but American viewers, perhaps expecting something else, refused to stop and read, to make the necessary commitment. Elsewhere, in other exhibits, Eames used all the tricks now accepted by contemporary exhibitors: bits of film, slide shows,

music, artifacts. In the "Franklin and Jefferson" show his ideas had to be read to be understood. The printed word dominated, print substituting for visual imagery, and Americans were disappointed, perhaps because the exhibit appeared in art museums where visitors expect their images to be visual. Europeans took to it more readily. In Europe, where it appeared in Paris, Warsaw, and London, viewers commited themselves to penetrating the print.

Many Eames exhibits explained science or scientists ("Computer Wall," "Copernicus," "On the Shoulders of Giants," "Newton"), and Eames liked working with scientists, recognizing in the way they worked—taking in masses of information and organizing it in manageable sections, then discovering relationships among sections— the very way *he* approached a problem. He would sift through a new sub-

"The World of Franklin and Jefferson," an American Revolution Bicentennial Administration exhibit designed through a grant from IBM Corporation and with the cooperation of the Metropolitan Museum of Art in New York, 1976.

ject, learning as much as he could, picking facets that seemed to him important, then transmitting these facets into plywood, onto film, or into the framework of an exhibit. In designing anything Eames leaned on his staff, asking everyone to think about a problem, expecting them to. He didn't like whistling in the shop; if you were whistling you weren't thinking.

Eames would approach anyone who could add even a drop to the pool of information he was collecting. His intensity as a researcher was part of his unceasing quest for perfection, and neither the intensity nor the perfectionism were tolerated easily by everyone who worked with him. Many left the office, and some remain bitter about the experience, but for those who stayed and adjusted, or who shared the same values, the Eameses studio was an exciting place to work. With the right colleagues Eames was a stimulating collaborator, and he was confident enough of this collaborative gift that he could make light of it.

Eames told Owen Gingerich, the Harvard astronomer, this about his collaboration with Saarinen on the first plywood furniture:

"It used to be this way: in the middle of the night, at three o'clock in the morning—this has something to do with what is known as the "creative process"—out of despair I would say something like, 'If we can't do anything else, let's paint it brown.' And then Eero would say, 'What? Of course we'll make it round. It's obvious that's the solution. Why is it that I'm an idiot? I never thought of it. You take it and see it should be round.'

"And I say, 'Wait a minute, you're mistaken. I just said a dumb thing about it being brown.'

" 'You're always trying to make me feel good. I never have any ideas; you

"Physics for a Moving Earth," an exhibition about Isaac Newton, designed for IBM Corporation, 1973.

A SEQUENCE OF 20th CENTURY IDEAS, EVENTS, AND ARTIFACTS

1900   1910   1920

"A Computer Perspective," exhibit for IBM Corporation, 1971.

have all the ideas. Why didn't I think of it?'

" 'But I said brown.'

" 'Don't give me that crap.' "

Rarely did Eames work on a project that he couldn't take into his shop and mold in his own way at his own pace. But once he ventured out. As an industrial designer, he entered for the first time the world of corporate engineering, asked by Stephens Trusonic, Inc., to give their Quadreflex high-fidelity speaker a look of newness and excellence. Eames, of course, brought to the problem of music reproduction no more knowledge than that of a moderately interested amateur. He began by sitting down with the engineers and asking questions: "What is the grille cloth for anyway?"

"To keep people from poking their fingers through the speaker cone."

"Why do you have to put a speaker in a box like this?"

"Maybe you don't."

Eames became a catalyst. He started the engineers thinking about the theory of speaker structure. Talking about the experience not long afterward, Eames said he regarded the process he started as his reward for taking the project. "The most grat-

ifying part of the whole project is the way the engineers took to the new ideas and developed them. We would say, 'Why not do it *this* way?' And they would say, 'Well, you *could*. But if you did then you'd have to move the . . . say, maybe that's not a bad idea. Look.' . . . And then they'd go to work on it. Their contribution was by far the greatest. What we did was second best." Eames called himself a "random element" in the process that gave Stephens Trusonic Quadreflex speaker, in 1956, a new look. It came out raised, floating, on an aluminum rod anchored to a four-footed, cast-aluminum base, much like the base for his lounge chair's footstool, which appeared the same year. The speaker's grille cloth—a colored, woven, saran cloth, stretched over an aluminum hoop—snaps off for cleaning. The cloth covers the round speaker, which dominates like a giant, fixed eye, the center of the square speaker box.

Eames saw engineers, designers, scientists, artists, poets all working with the same process, one which seeks to discover and articulate relationships. The process must, he realized, eschew dogma and self-expression. In a particularly revealing statement Eames once said, "The preoccupation with self-expression is no more appropriate to the world of art than it is to the world of surgery. This does not mean I would reduce self-expression to zero. I'm sure that the really great surgeons operate on the edge of intuition. But the rigorous constraints in surgery—those are important in my art."

But if Eames relied on precise measures and procedures in his search, his approach was nonetheless intuitive. He saw, as Kepler, Darwin, and Einstein have pointed out, that intuition is the secret. His distrust of self-expression grew out of his realization that quirks of personality can destroy a design's solution. A designer must work through his own prejudices and quirks, Eames once explained, until he can design beyond them.

Eames also decried the canons of taste to which so many designers consciously conform. His search for relationships and structures led him far beyond the design fads that fattened and faltered following World War II. If the bulbous, tail-finned 50s, the pop 60s, and the postmodernist 70s all stemmed from disillusion with machines capable of turning on us the horrible, obliterating mechanisms of war, then the Eameses' vision and intuition, their handcrafted approach, carried them through to a purer expression. It is why their designs have no dogmatic look or style.

The Eameses viewed their work as "lifelong learning" as "progression toward breaking down the barriers between fields of learning . . . toward making people a little more intuitive." Their work was both a way of seeing and a way of making others see.

Buckminster Fuller, who sees many things more clearly than most, called Eames an artist-scientist. He had, says Fuller, "a beautiful lens."

This is the twenty-fifth year in which *Industrial Design* magazine has presented a review of the year's industrial design. Looking back 25 years at the first review, one finds examples of the type of design, the kind of seeing that Eames sought. In fact an Eames design appeared in that 1954 review: the sofa he did for Herman Miller, a high-backed, chrome-legged, collapsible sofa. And it looks as right now as it did then. The Eameses used one in their living room, covered much of the time with an animal skin and lots of colored pillows.

Stevens Trusonic Quadreflex speaker, 1956.

Sofa compact, Design Review award winner, 1954. Photograph courtesy Carl Ruff Associates.

Other designs that year achieved the same clarity of form: the dinnerware George Nelson designed for the Pro-Phy-Lac-Tic Brush Co.; a 150-lb fire extinguisher on wheels, done by Raymond Loewy for the Ansul Chemical Co.; the near perfect Keyhole Saw that Garth Huxtable designed for the Miller Falls Co.; and the Circular Vault Door Henry Dreyfuss did for the Fifth Avenue window of Manufacturers Trust, a Skidmore, Owings & Merrill building, in New York.

"Steamlining" was still a word used by designers, though it was passing as we entered the breastlike bulge and tail-fin era. It is perhaps surprising that few of the items in the review showed either bulge or streamline, though the Aquaped that G. McRoberts and C. A. Gongwer designed for Aerojet General had them, and so did the C/F fountain pen Harley Earl, Inc. did for Waterman Pen.

Even in 1954, when industrial design and designers were still finding their way (more than they are today), good design showed a timelessness that comes only from clear thought and undistracted vision, from an understanding of structure and a sensitivity to relationships of parts. Some of the men responsible for founding the profession were represented—Loewy, Earl, Dreyfuss. Most of them are gone now, and so is Eames, who came later. Almost all the faces in this year's review have appeared since then. Robert P. Gersin & Associates, who have nine items in this year's review—the most of any single designer or firm—opened their office in 1959.

The statement of purpose made by the editors of that first review remains remarkably apt, proving perhaps that good editing, like good design, must have clarity of vision and statement and a dollop of intuition. It read:

"This review is not, by a long shot, a complete coverage of every good product put on the American market in 1954. If we could have found them all, they would have filled the pages of an encyclopedia. Nor is it a collection of immortal design—they would have made a pretty skimpy issue. It is simply a group of several hundred products selected from material submitted or sleuthed that the editors feel have real design merit."

If you change 1954 to 1978 and select the material by jury, not mere editorial consent, you have this year's annual *Design Review*.

CONSUMER PRODUCTS

Cuisinart Triomphe 500 Food Processor, Commercial
Corning Ware® MC-1 and MC-2 Fast Food Dishes
Thermos "Touch Top" Model 2647 Beverage Dispenser®
Clairol Travel Hairdryer, "1 for the Road"
Copal World Timer Alarm Clock
Texas Instruments Electronic Analog Chronograph Watch, Models
851 and 852
Texas Instruments "Speak & Spell" Talking Learning Aid
J.C. Penney Hand-Held Vacuum Cleaner
McCulloch Pro Mac 610 (Farmer) Chain Saw
Prototype Aqua-Touch Water Faucet Attachment
Prototype Hand-Held Hairdryer
Prototype Backpack with Cot

Arthur W. Ellsworth
Vice-president of Design West in Newport Beach, California, Arthur Ellsworth studied engineering at the University of Minnesota and industrial design at the Art Center College of Design in Los Angeles. He did product design and development work for Lewis and Tweedie Industrial Design, then moved to Design West when it was formed in 1962. His responsibilities include overall operations management and coordination of all design and model development.

Cletus Durcan Hudson
The consumer product jury's consumer is Cletus Durcan Hudson. Her product experience comes solely from having bought and used, with her family, the types of products reviewed by the jury. She has a master's degree in social work from the University of St. Louis and has worked extensively with social services in the pediatric departments of hospitals in St. Louis and New York. For two years she supervised Bellevue's department of social services, pediatrics. While there she initiated and conducted a study of abused children. Besides raising a family and renovating an old house, Hudson is active in community affairs and has worked for small companies, developing multimedia educational material.

Richard Montmeat
Since 1976 Richard Montmeat has been manager of General Electric's industrial design operation in the company's major appliance and marketing division at the Applied Research & Design Center in Louisville, Kentucky. He has been with GE since 1948, a year after he graduated from Pratt Institute, working on electronics, radio and TV, consumer electronics, and housewares.

Jerrold W. Ross
Jerrold Ross is vice-president of marketing and new product development for Dansk International Designs Ltd., in Mt. Kisco, New York. He went to Dansk from Pratt Institute where, as an associate professor, he was director of the design division. Before that he was president and creative director of Synergy Design, Ltd., a firm specializing in international marketing and development of new consumer products.

E.B. White once noted that "A home is like a reservoir equipped with a check valve: the valve permits influx but prevents outflow. Acquisition goes on night and day—smoothly, subtly, imperceptibly.... Under ordinary circumstances, the only stuff that leaves a home is paper trash and garbage, everything else stays on and digs in."

If industrial designers are not directly responsible for this inflow, they *do* influence it—"smoothly, subtly, imperceptibly"—and encourage it openly by increasing the lure products have for consumers. What is that lure? What gives one version of a product the special appeal that will send it through the check valve of a home and keep it there? Jurors Arthur W. Ellsworth, Cletus Durcan Hudson, Richard Montmeat, and Jerrold W. Ross concluded that much of what they saw during the day they spent considering consumer product submissions for this review was well designed—easy to look at, use, and manufacture. But in a society where the level and awareness of design are rising, where TV sets and toasters are almost universally well designed, design may not be enough to fire acquisitiveness (though poor design will certainly obstruct it).

Given this situation, the jurors decided incisively, with little discussion, what to look for. While recognizing that an obvious and important element of design is aesthetics, they refused to confuse aesthetics with cosmetics. Making an outdoor grill red this year instead of black or giving a Thermos jug a rainbow-patterned facade may offer these items momentary security in the eye of the consumer or may trigger a fleeting lust for possession. But the jurors rejected such cosmetics as temporary and faddish. What is needed, the jury felt, to secure a product a place in these pages—and perhaps in a home—is innovation. The jury searched for innovation, seeking new shapes and performances the way a football halfback seeks daylight. Time and again the jurors rejected products that, although handsomely designed, represented what they called "the state of the art." If a chicken basket, an ice cream maker, a hi-fidelity speaker or turn table, a hair dryer, a freezer, an espresso maker, a telephone, a sewing machine, or any of the 101 products and prototypes the jury reviewed represented only "the state of the art," it was cast aside.

What the jurors sought was either a new form for an old product or a technological wrinkle giving an old product a new use. Rarest, and most eagerly sought, was an entirely new product, one with both new form *and* new use. If a power tool manufacturer can design a new tool, he will capture the attention and channel the acquisitive drive of the home workshop buff. This manufacturer also has a chance to try a new form, but "extracting design from a product's marketing program is a difficult task," one juror reminded the group. "Design is only one of so many other considerations." Still the jury looked carefully for the aesthetic form that sets one well-designed product apart from its peers.

What emerged from their search was a stack of only nine products and three prototypes—items that show innovation in what they do, how they do it, or how they look.

Appearance was perhaps the most important aspect of the jury's decisions. If an item had an innovative look, as the Copal alarm clock (page 56) and the J C Penney's hand-held vacuum cleaner (page 60) do, the jurors immediately warmed to it. If an item showed innovation in mechanism coupled with an ordinary appearance, as Texas Instruments' electronic watch does, their enthusiasm was genuine but muted. Grandmother might not have known what to make of the Copal alarm clock, while feeling almost immediately comfortable with the Texas Instruments' electronic watch. And these considerations are, of course, the point of these designs. The Copal alarm clock, though containing old technology, is meant to appeal through a seeming improvement in its form, like a woman who speaks in cliches but dresses charmingly ahead of the fashion. The Texas Instruments' watch, with its revolutionary technology, was meant to look comfortably familiar. It represents an electronics firm's first foray into the time marketplace, and the watch must appeal as widely as possible.

The item that probably most excited the jury was also designed by Texas Instruments. It is what that company calls a "talking learning aid" (page 59)—a small, portable, keyboard device whose solid state circuitry talks to children, saying things like, "Spell farm." And "That's right" or "No, that's wrong; try again." The device represents innovation and was embraced by the jury.

If the talking learning aid sounds, from this brief mention, as if it is more gimmick than aid, it is not. The jury shunned gimmickry. Their rejection of the Bell Labs Noteworthy phone—a Trimline telephone attached to a phonebook holder whose front surface is a built-in corkboard or blackboard, designed by Henry Dreyfuss Associates—was precisely because the jurors thought the phone more gimmick than innovation. It looks like a "Mickey Mouse phone" was one comment, if not a generally shared perception. The jurors decided that a telephone with a corkboard selling for $83 did not represent a better design of 1978.

They rejected, too, an egg scrambler, designed by the Stansbury Company for Ronco Teleproducts Inc. of Elk Grove Village, Illinois. This device holds a bent pin, which when stuck through an egg shell rotates electrically scrambling the egg inside its shell before the egg is broken into a pan. All this, the jury thought, failed to fulfill their image of innovation. The Ritepoint Company's Pyramid Desk Pen, designed by Intelplex, came close, with its conical body that lets the pen stand upright, announcing its position on a cluttered desk. "It will sell," proclaimed the jury as they passed it by.

If the talking learning aid had a weakness or was after all less than an ideal, it was because of its appearance. Though jurors did not rave about that appearance, they judged it good for its market, while they felt Polaroid's Sonar One Step, which they rejected, fell short aesthetically. "An aesthetic mistake," one juror called it, though he and his fellows applauded the camera, which uses sonar to adjust its distance setting, as an "important technological improvement."

In 1978's product design two trends are discernible. One is toward speed and efficiency, the other rejuvenates an old friend: status. The design of many products today is directed toward making you, the user, look like a professional, while putting out minimum effort. Perhaps indicative of both trends (speed and status) is the Henry Dreyfuss–designed Singer Touch Tronic 2001 memory sewing machine, which the jury considered but rejected. With a retail price of $900, the machine feeds two layers of material beneath the needle. An operator can select any of 27 preprogrammed stitches merely by touching a picture of the stitch on the housing. The machine is, the jurors felt, "a marketing decision not a design."

Speed and efficiency are offered by the Cuisinart food processor (page 50), by Corning's fast food dishes (page 52), and less dramatically by Thermos's "Touch Top" server (page 53). One theory has it that Americans, surrounded by complexities, living in households supported by two or more workers commuting long distances to offices and factories, want to spend less time on home chores. So gadgets that make home life less complex (or seem to) are eagerly sought. The catch may be that we spend the time the gadget saves taking care of the gadget. We are at the stage where feelings are mixed. Some consumers want simpler products, things that break down less frequently. Others are willing to pay for convenience and are less concerned about potential breakdown.

The jury discussed one further consideration. J C Penney's is paying close attention to design, redesigning the products it sells, the atmosphere it sells them in, and the sales materials that promote them. The question the jury raised is whether Penney's design goes beyond the acceptance of its clientele, whether the new image will appeal to the shoppers the company has traditionally attracted, or whether it will have to rely on new ones. Whatever is happening, Penney's profits for 1978 were up, and the jury applauded the approach, if not every detail, of the company's design program. They singled out only one Penney's item, its portable vacuum cleaner. But then the jury was looking for innovation, as well as good design, for items that make you want to take them through the check valve of your home and let them "stay on and dig in."

Noteworthy Telephone, Design Line Series. Client: Western Electric Co., New York; manufacturer: Bell Laboratories, Indianapolis, Indiana; consultant design: Henry Dreyfuss Associates, New York: Donald M. Genaro, Jack McGarvey, Alvin R. Tilley.

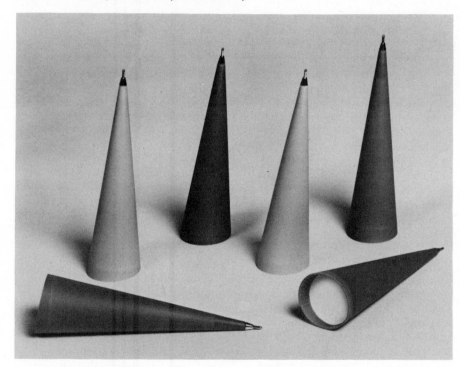

Pyramid Desk Pen. Manufacturer: Ritepoint Company, Fenton, Missouri; staff design: Alan E. Sherman, Wells S. Bearinger; consultant design: Intelplex, Maryland Heights, Missouri: Harry Rosenberg, Harold Koeln.

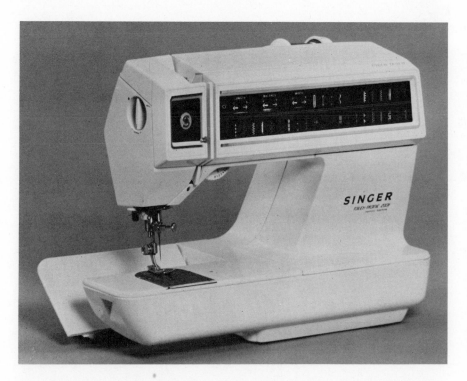

Singer Touch-Tronic 2001 Memory Machine. Manufacturer: The Singer Company, Elizabeth, New Jersey; staff design: John Van Duyne; consultant design: Henry Dreyfuss Associates, New York: Donald M. Genaro, Chris Felix.

Polaroid SX-70 SE Land Camera, SONAR OneStep. Manufacturer: Polaroid Corporation, Cambridge, Massachusetts; consultant design: Henry Dreyfuss Associates, New York: James M. Conner, James M. Ryan.

Cuisinart aims its redesigned food processor (with a larger, quieter motor) at anyone preparing large meals, whether a housewife cooking a family Thanksgiving dinner or a chef preparing a Rotarian banquet. The machine, like its predecessors, will cut, slice, mix, puree, shred, knead, grate, or pulverize, depending on the knives and discs one sets into it and the setting one gives them. Moreover, it will perform these tasks in what seems a blink of an eye. Engineering improvements include braking circuits that will stop a rotating blade within four seconds after you twist the cover to turn the unit off. Only when the blade stops can the cover be removed. The designers changed the base configuration to accommodate a drop-on cutting board and to provide a surface for graphics. They also changed the unit's proportions. It measures 18 x 10½ x 15⅜" (45.7 x 26.7 x 39 cm). Retail price: $600.

Materials and Fabrication: base, top cover, and motor support: aluminum (sand castings); work bowl and cover: tinted polycarbonate; cutting discs and blades: stainless steel; cutting board: laminated rock marble (machined). Base is painted white with silkscreened graphics. Work bowl and cover are tinted bronze.

Manufacturer: Cuisinarts Inc., Stamford, Connecticut.

Staff Design: Carl Sontheimer, chairman of the board, president; Al Finesman, director of the product development; Robot-Coupe, France, engineering and design staff.

Consultant Design: Marc Harrison Associates; Marc Harrison, chief designer, president.

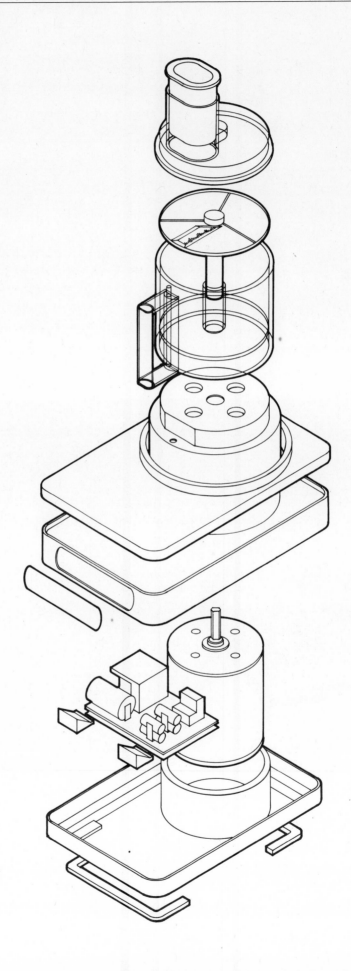

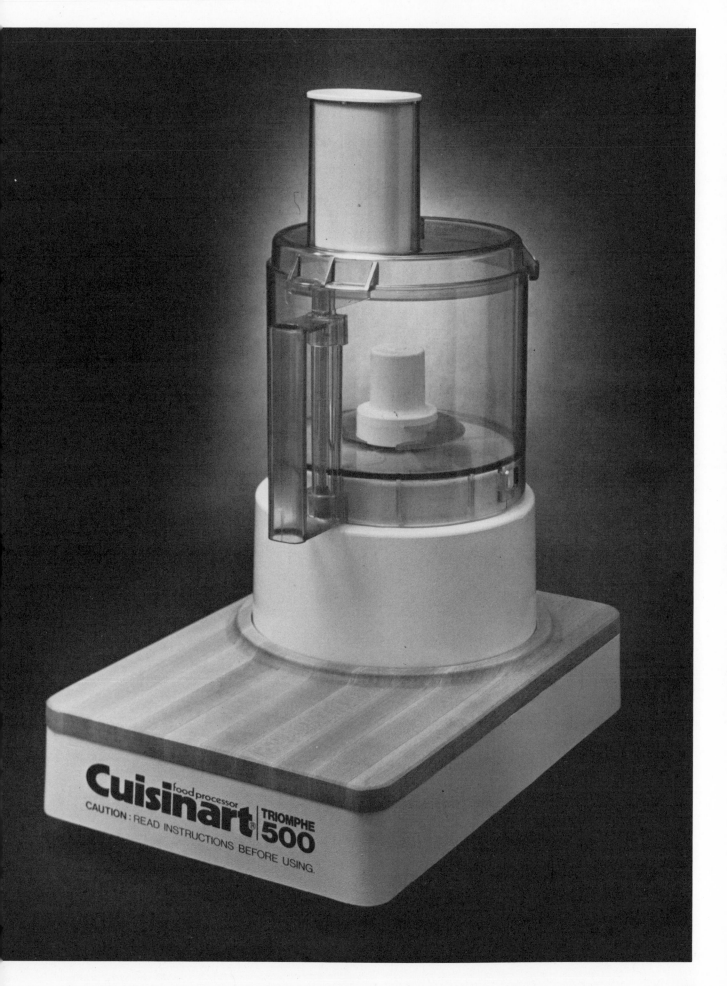

In microwave cooking it's important that the cooking container fit the food snugly, cupping its juices close to the cooking food. Corning Glass Works' new Corning Ware Fast Food Dishes are so designed to hold standard 1- or 2-pound (9 kilograms) frozen food blocks. A cook merely unwraps the frozen food package, slips the food block into the Fast Food Dish, and pops both into an oven—conventional or microwave. For cooking the dishes have glass lids, and for freezing each dish comes with a companion polyethylene cover. In this way the dishes can be used by anyone who freezes his or her own food. The dishes become a mold, holding the food as it freezes. When the food's frozen, you take it out of the dish and wrap it for freezer storage. To cook it you put it back in the dish, cover it with the glass top, and put the whole thing in the oven.

As an additional feature each dish has four knobbed, bottom feet which fit into the glass cover's matrix of circular, symmetrical indentations. So the cover can become a trivet to protect table surfaces. The MC-1 dish, which measures 7 x 5½ x 2" (17.8 x 14 x 5.1 cm), retails for $8.99. The MC-2, measuring 10½ x 6¾ x 2" (26.7 x 17.1 x 5.1 cm), retails for $12.99.

Materials and Fabrication: dish: Corning Ware Ceramic (pressed); glass cover: Pyrex soda lime tempered glass (pressed); plastic cover: U.S.I. Chemical petrothene NA-270 (injection molded).

Manufacturer: Corning Glass Works, Corning, New York.

Staff Design: P. Dennis Younge, senior product designer; Larry W. Edwards, senior product engineer; George A. Neil, marketing manager.

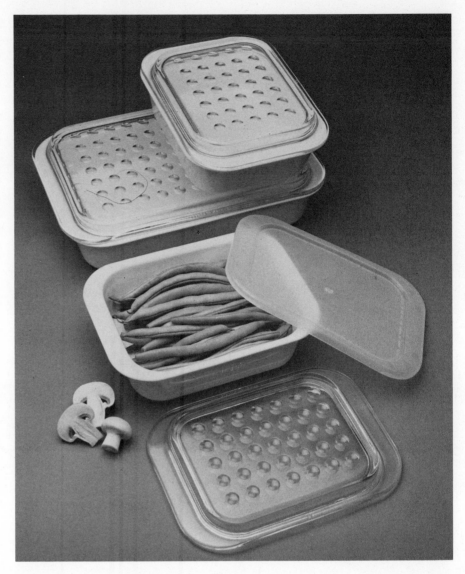

Thermos's model 2647 will keep up to 2 liters of liquid hot or cold, dispensing them when you press a button, or, more precisely, its top. Not only does the internal pump make filling a glass from this vacuum bottle easy, but it also means the top can be kept in place, allowing the liquid to retain its desired temperature longer. Pump assembly and spout are designed to be taken apart and cleaned under a water faucet. The *Design Review* jurors liked the model 2647's proportions, its clean lines, its unobtrusive tote handle, and its innovative pump serving. (Though a similar unit is marketed in Japan, this was the first to appear in the U.S.) Height is 12" (30.1 cm); top diameter: 5⅛" (13 cm); bottom diameter: 6⅜" (16.1 cm). Suggested retail price: $19.95.

Materials and Fabrication: pump cover, handle ring, and pump-spout: ABS polymer (injection molded); case, handle, and base: polypropylene (injection molded); pump assembly: polyethylene. All model 2647 units are ivory colored with one variation, the Autumn Harvest, which is decorated with undulating colored bands applied by offset printing.

Manufacturer: Thermos Division of King-Seeley Thermos Co., Norwich, Connecticut.

Staff Design: Richard A. Tarozzi, industrial design; George W. Fuller, product engineering; Edward C. Carlson, product development engineering; Carl L. Johnson, product drafting.

Consultant Design: Group Four Inc.; Frank von Holzhausen, graphic design.

Clairol calls its compact travel dryer "1 for the road" and means it to go places. A switch on the main body converts the dryer for use with 220-volt current found overseas, and the unit's size, roughly a 4 x 4" (10.2 by 10.2 cm) package with the handle folded up, takes little space in a travel bag. By pivoting out a built-in stand in the folding handle's back side, one can set the dryer down and angle it for hands-off hair or clothes drying. A slide switch on the main body provides three blower speeds and three heat levels. The dryer's finish is a high gloss orange; graphics, pivot cap grill, and cord are black. Suggested retail price: $22. "Graphically and visually very attractive," said the jury.

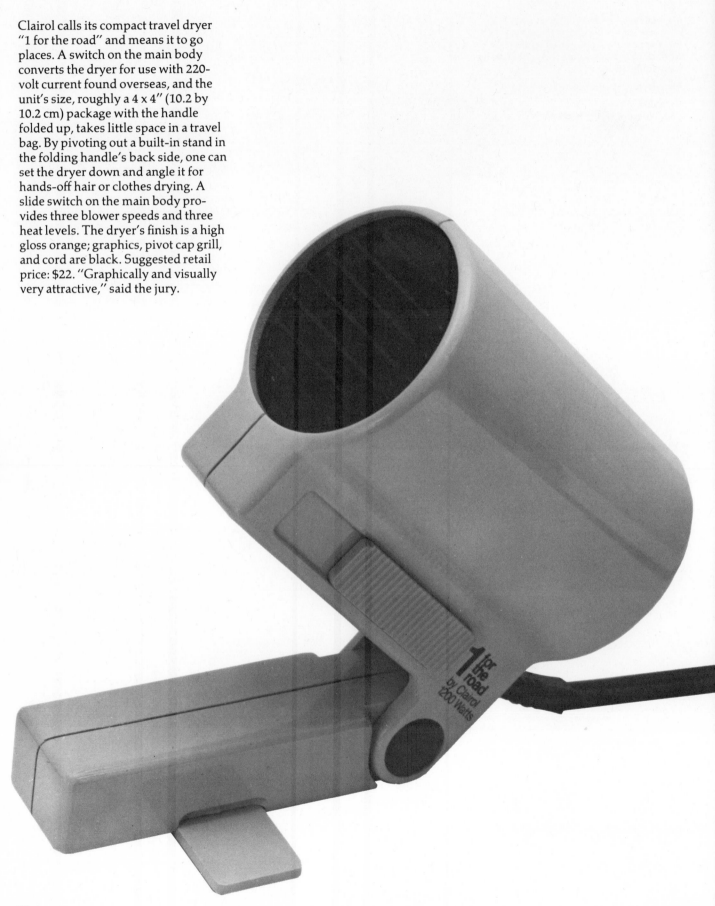

Materials and Fabrication: dryer's main body: polycarbonate (molded in two halves and held together by the handle, a forward screw and top-locking sections; handle: polycarbonate (ultrasonically welded).

Manufacturer: Clairol, New York.

Consultant Design: John Wistrand Design: John Wistrand, R.W. Fyfield.

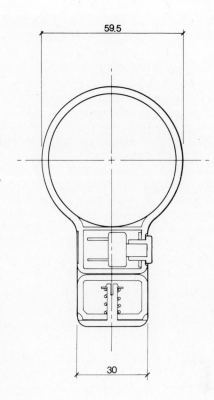

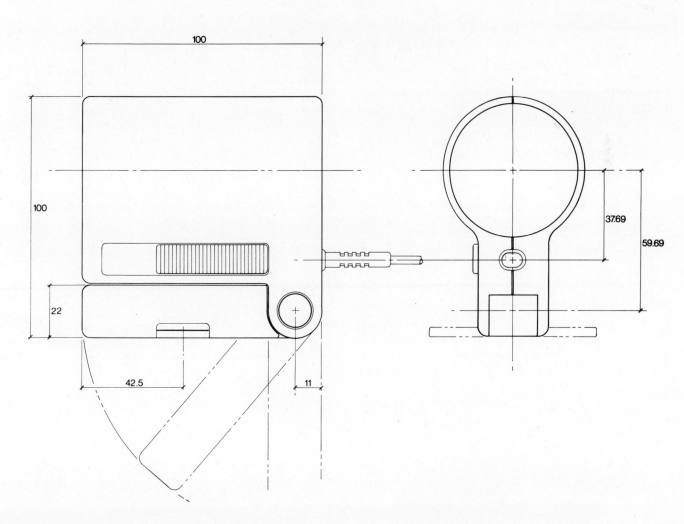

Copal World Timer Alarm Clock

Designed for the globe-trotting traveler, the World Timer Alarm Clock (by Japan's Copal Company Ltd.) is intended to convey quality and sophistication to a sophisticated buyer; it sells for $29.95. With its introduction Copal hopes to build prestige in the international market. What the clock offers is easy time adjustment as a traveler passes through world time zones. The designers positioned the controls so as one holds the compact clock [2⅞ x 4¾ x 1⅛" (7.3 x 12.1 x 2.8 cm)] in a palm, the thumb can rotate a knob that selects any one of 24 major

cities, each in one of the major time zones. Turning the time zone control automatically adjusts the time. (One juror suggested wryly that the clock would be a help to a traveler who on waking has forgotten what city he's in.) To the left of the time display two thumb wheels control the alarm setting. A slide switch, also on the left, sets the alarm and turns it off. Power comes from one C-cell battery. A movable cover at the bottom of the clock's front side hides time setting controls. The jury applauded the clock's innovative appearance.

Materials and Fabrication: main frame and side panels are ABS polymer (injection molded) with a sandblast texture; front, top, and back surfaces and lower front section which opens to expose time set controls and battery: .8 mm aluminum.

Manufacturer: Copal Company Ltd., Tokyo, Japan.

Staff Design: Kiyohisa Yabu, product manager, sales division; Kiyoyuki Arai, senior engineer, mechanical development; S. Onishi, senior engineer.

Consultant Design: Plumb Design Group: Andrew T. Serbinski, associate.

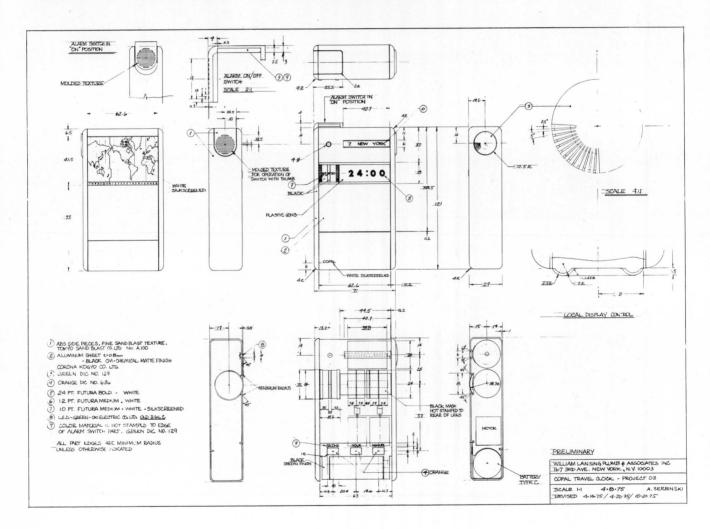

Though the *Design Review* jurors felt Texas Instruments could improve the aesthetics of its electronic analog chronograph, they applauded what the watch does and how it does it. Without moving parts, without springs, gears, wheels, arms, switches, of conventional mechanical hands, the watch measures and displays time. It uses, instead, electronic circuits and liquid crystal display indicators, which let you know what time it is in hours and minutes by pointing to the traditional markers on the watch face. "It is an answer to the problems of the digital watch," said one of the jurors. And the electronic analog will do more than just tell the hours and minutes. At the push of the button the dial converts to a display of minutes and seconds. Moreover, the watch will keep track of elapsed time to the tenth of a second for as long as 12 hours. The press of the elapsed time button displays elapsed time in hours and minutes. Pressing it again gives a display in seconds and tenths of seconds. And though you may shift back to normal time display, the elapsed time function will continue. You can set the watch to display time in a different time zone, then have the watch display time in either zone by merely pressing its buttons. If you press the main button twice, the watch will display (for two seconds) the day of the week and the date. The watch calendar adjusts automatically for long and short months.

"They've taken the newest technology and put it to work to operate the way people use watches. That to me is significant," said one juror.

In its square version the watch case measures 1.4 x 1.4 x .365" (3.5 x 3.5 x .9 cm). As a round watch it measures 1.4" (3.5 cm) in diameter. Suggested retail price is $250 to $325. Though its face is readable, the jurors had reservations about the watch's overall appearance. "It's designed like jewelry," was one comment. That the design was not cleaner and simpler, that it was "busied up," that it looked "like a lot of stuff" were the jury's objections. "Technologically it's significant," said a juror in conclusion. "We hope the future will bring better design."

Materials and Fabrication: case: either stainless steel or gold-plated brass (forging); crystal: mineral glass with vacuum metalized decoration.

Manufacturer: Texas Instruments Inc., Lubbock, Texas.

Staff Design: E.J. Sulek, design manager/director; E.J. Sulek, W.J. Lawrence, industrial designers; Robert Noble, integrated circuit design; Bob Gruebel, program manager; Burt Marks, display; Gene Dahl, case design; John Marcotte, Frits Kuyt, module design.

With "Speak & Spell," Texas Instruments puts solid state circuitry to work helping children learn to spell. The device talks to the child, telling him or her which of 200 preprogrammed words to spell. "Spell house," the machine will say. The child punches out the word on the keyboard; the tiny machine shows his spelling of the word on a display screen and either praises him, if he's right, or asks him to try again, if wrong. The device has no moving parts, no records or tape recorders commonly used in talking toys. Within the device a single integrated circuit, smaller than an asprin, carries the digitally coded data needed to produce synthetic speech electronically.

Measuring 10 x 7 x 1.3" (25.4 x 17.8 x 3.3 cm) and weighing 1.2 pounds (534 grams), the unit can be carried around by its molded-into-the-case handle. This handle gives a child (a typical user is between 7 and 12 years old) something to hang onto, reducing his chances of dropping the machine. Suggested power comes from four C-cell batteries.

And the machines is versatile. At the touch of a button the machine will play games with you. One, for instance, is a variation of Hangman, in which a player guesses words one letter at a time. When the game is over, the machine announces either "You win" or "I win."

"Technically," said one juror, speaking of how they all felt, "it is a sensational product." No one, however, praised the Speak and Spell's appearance, though they noted the rounded corners, the handle, and the appeal the graphics should have for children. Said one juror, "It appears to be a well-thought-out product."

Materials and Fabrication: case: ABS polymer (injection molded); living hinge keys are overlayed with aluminum, with lithographed color and graphics.

Manufacturer: Texas Instruments Inc., Lubbock, Texas.

Staff Design: E.J. Sulek, design manager/director; W.J. Lawrence, R. Nesbitt, industrial designers; L. Brantingham, electrical engineer, circuit design; G. Frantz, electrical engineer, program manager; R. Chang, W.R. Hawkins, mechanical engineers, case design.

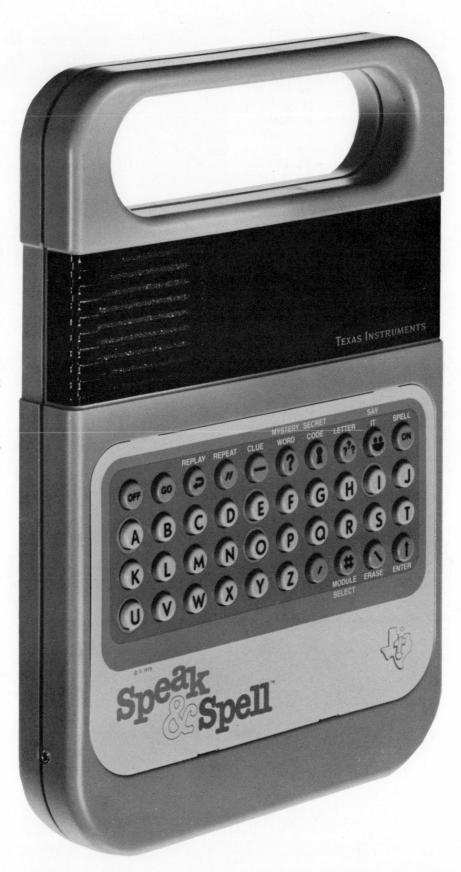

J C Penney Hand-Held Vacuum Cleaner

J C Penney's redesign of its hand-held vacuum cleaner gives it a versatility not usually found in small, hand-held machines meant for cleaning cars, workshops, boats, or awkward spaces in the home. Penney's model comes with a shoulder strap, a molded-in handle, and housing storage space for attachments—the 4' (1.2 m) flexible hose and an 18' (5.5 m) cord.

All this weighs 9.2 pounds (4.2 kilograms), including the cleaning tools. The housing measures a compact 12 x 12¾ x 7'' (30.5 x 32.4 x 17.8 cm). There is a reusable cloth bag and a 1 hp motor. The designers specified identical screw-together halves, saving, they claim, tooling costs.

The jurors were not wholehearted in their support for the portable vacuum. They thought, for instance, that the open spaces in its housing where cord, tools, and hose are shoved for storage might be inadequate and that the machine might begin to look messy, with cord and tools hanging out of it, once it had been used a couple of times. They felt, too, that the housing's upper edge might be sharp. But they concluded that in form, versatility, and price ($44.99) the vacuum represents a departure to be encouraged.

Materials and Fabrication: body: heat and impact resistant styrene (injection molded).

Manufacturer: Douglas, Akron, Ohio.

Client: J C Penney, New York.

Staff Design: for Douglas: Ken Meiser, executive vice-president, management and sales; Dave Colbert, Tim Wilson, engineers. For J.C. Penney: Frank Palazzolo, project manager; Ron Pierce, Jan Tribbey, Jeanne Kirsch, Chris Hacker, product designers.

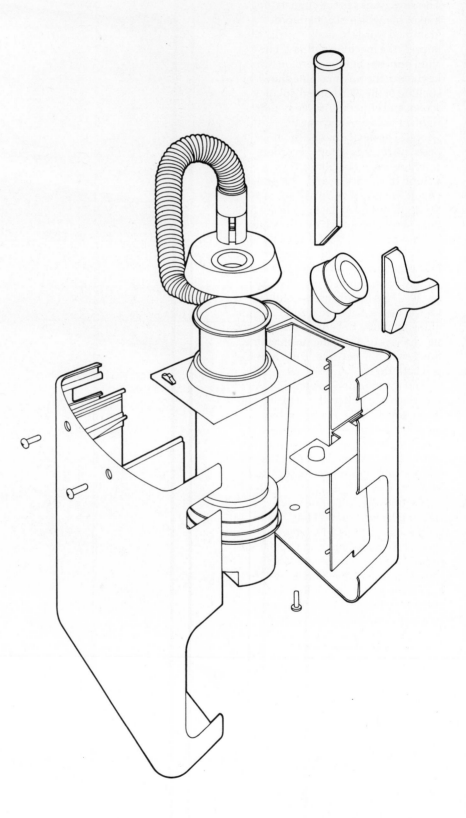

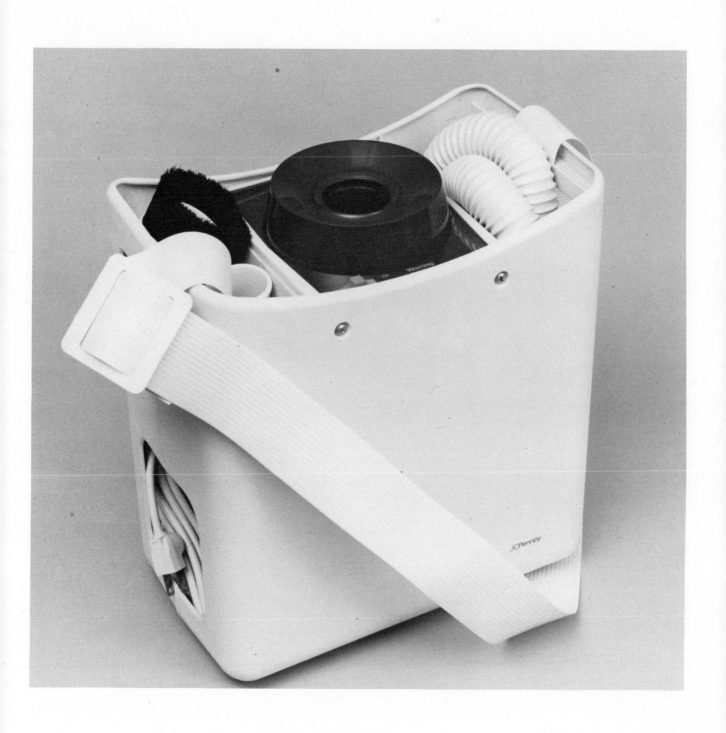

Suitable for a range of wood cutters—farmers clearing woodlots, suburbanites trimming view-blocking limbs, even the semi-professional cutters—McCulloch's 610 chain saw has features not all its competitors have in a comparable saw. Besides, McCulloch says, the 610 retails for 10 to 30 percent beneath its competition's prices. Among the designed-in conveniences and safeguards are a chain brake, a hand guard, a handle that absorbs engine and cutting vibration, an electronic ignition, a fuel tank sight gauge, a muffler shield, and a catcher that will grab the chain should it accidentally break. By using foamed-glass-filled nylon for the handle, the designers cut manufacturing costs, they claim, and besides made the handle stronger, easier to grip (it has a molded-in texture), lighter, and capable of absorbing more vibration than previous McCulloch handles. Another cost reduction came from redesigning the 17.2-fluid ounce gas tank. It is now blow-molded from a single piece of polyethylene. They say the new tank is also tougher than the two-piece glued and sealed magnesium one they used before. And the new one has a molded-in fuel gauge. All the saw's controls are identified by international colors and graphics. The 610's housing is yellow; its handle, chain guide bar, and chain brake-hand guard are black.

The 3.7 cu in. (60 cu cm) two-cycle engine is said to produce more horsepower with less exhaust noise than previous engines McCulloch used. Weight is 17.7 pounds (8 kilograms) when the saw is used with a 16" (40.6 cm) bar and chain, and the suggested retail price is $229.95.

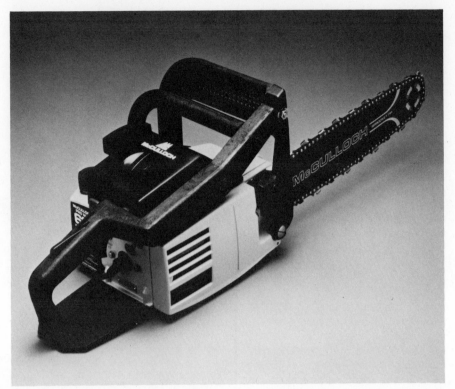

Materials and Fabrication: vibration dampened handle system: foamed-glass-filled nylon with rubber mounts; air cleaner cover and air silencer tubes: foamed-glass-filled polypropylene (injection molded); starter drum; mineral-filled nylon; gas tank: high-density polyethylene (blow molded); oil tank, clutch cover, fan housing, crankcase, bottom skid, cylinder shroud, and air box: AZ91B magnesium (die cast and machined); muffler: 4130 alloy steel (stamped and plated); trigger and throttle latch: foamed-glass-filled nylon.

Manufacturer: McCulloch Corporation, Los Angeles, California.

Staff Design: George E. Maffey, vice-president of engineering; Ronald W. Hottes, director of product development; John L. Zimmerer, senior project engineer.

Consultant Design: Tony Carsello Design: Tony Carsello, industrial designer.

Patented, but not yet manufactured or sold in the United States, Aqua-Touch will help save water, claim its designers. It will attach to any faucet threaded for an aerator (with an adaptor) and provide touch-control of waterflow at the spigot. By pointing the Aqua-Touch's chromed lever down into the sink, then pushing it back against spring pressure, with a hand, finger, toothbrush, cup, bucket, or glass, you start the waterflow. Releasing the pressure stops the flow, as the spring-loaded lever returns to the off positon. If you want a steady flow of water, you merely pull the lever up until it locks in place at right angles to the spigot. You still control water volume and temperature with the faucet's main controls. Designed to retail for about $7, the Aqua-Touch's appearance is partly determined by tooling costs. Its heavy draft angles, leading away from its midline, are meant to minimize tool wear during production. Overall dimensions: approximately 4 x 1½ x 1½" (10.2 x 3.8 x 3.8 cm).

Materials and Fabrication: main body and handle: chrome-plated zinc (die cast); one piece union joint adapter and snap-together valve stem: acetal.

Developer: Water-Plus, Los Altos, California.

Consultant Design: INOVA: David C. Anderson, president, industrial design, mechanical design, and project management; Justin L. Dantzler, designer, basic idea.

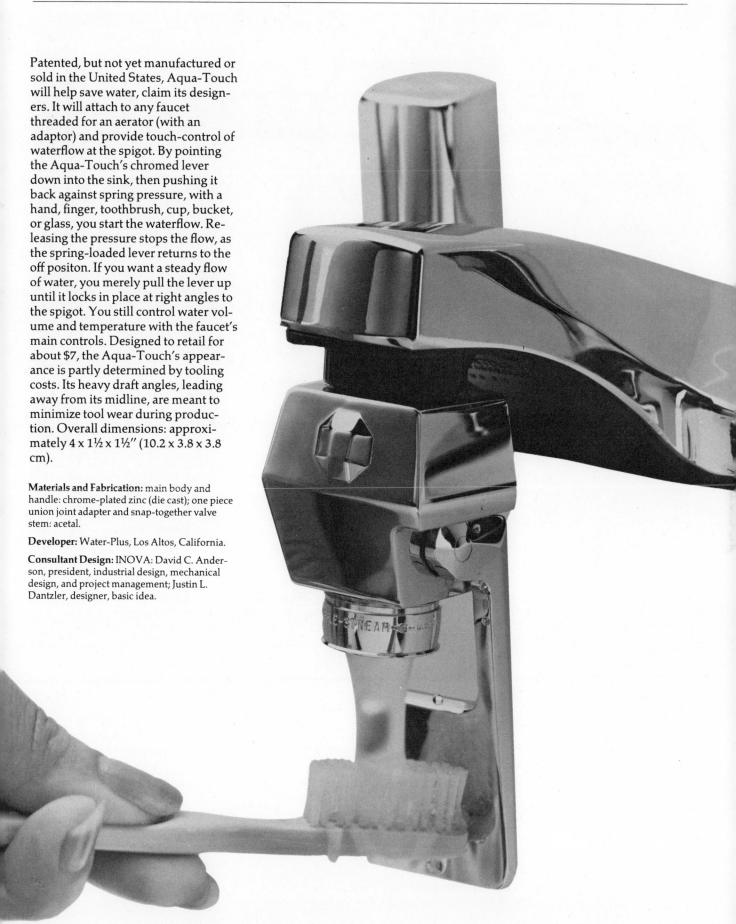

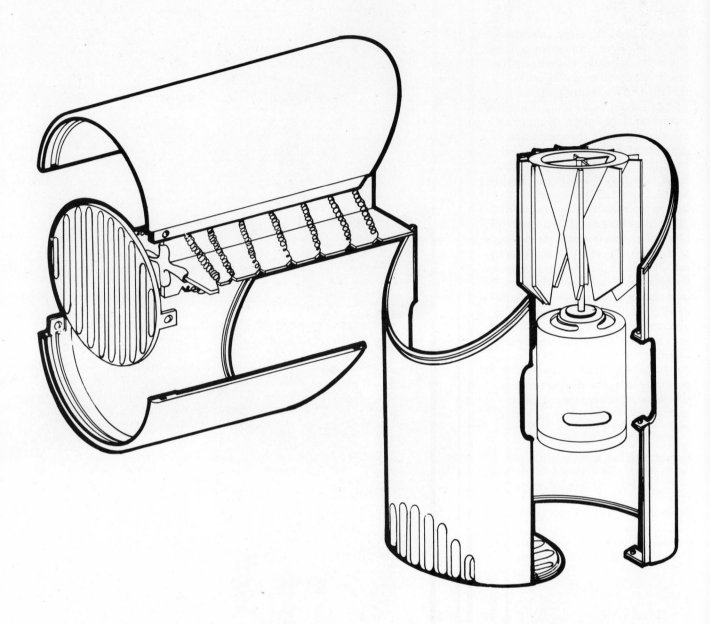

Imagine a hand-held hairdryer, tubular, with an elliptical cross section, a blower head that will pivot 350°, and a removable cord. That's what Marcie Lipschutz designed for a junior class project at the Philadelphia College of art. Schick, the personal care company, supplied the class with a consumer market survey and some older hairdryers that the class could analyze and dissemble. The problem was to provide optimum power in a compact, easy-to-store, traveling hairdryer, one that was lightweight, balanced to be self-supporting and easy to control, and one whose airflow could be di-rected accurately. Lipschutz's solution measures 6.75″ (17.1 cm) when the dryer is stretched out in an upright position for packing. It is 2″ (5.1 cm) wide. For drying the head pivots at right angles to the handle, forming an inverted L. Air intake covers the unit's entire bottom; the power switch is on the side beneath the user's thumb.

Materials and Fabrication: housing: ABS polymer (injection molded); snap fit assembly with self-tapping screws in bottom.

Designer: Marcie Lipschutz, under the direction of Bob Drobeck, product design instructor, Philadelphia College of Art.

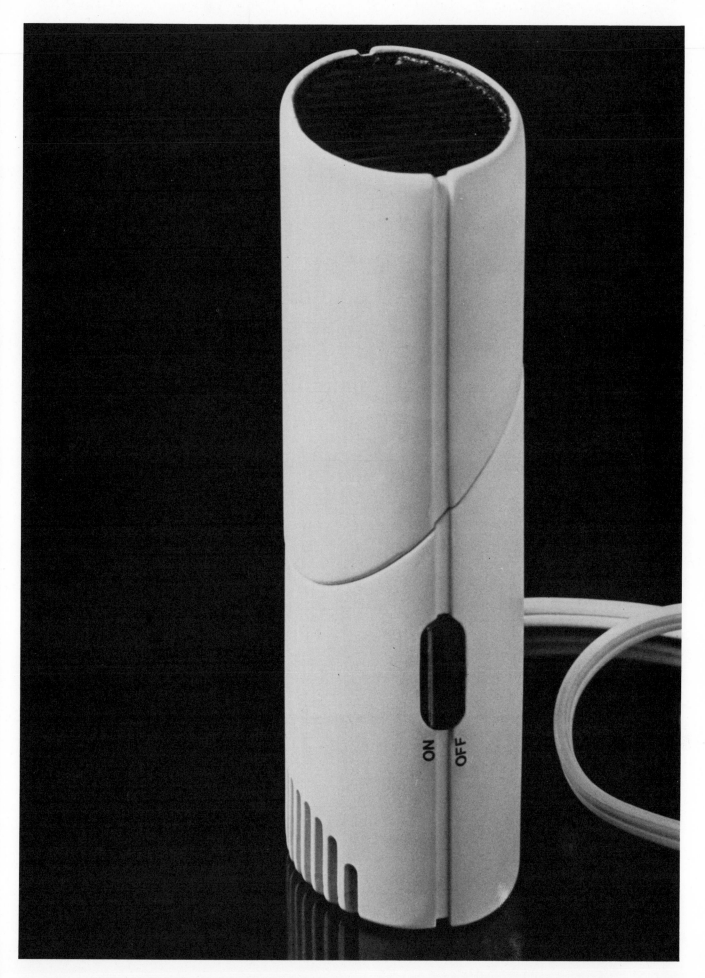

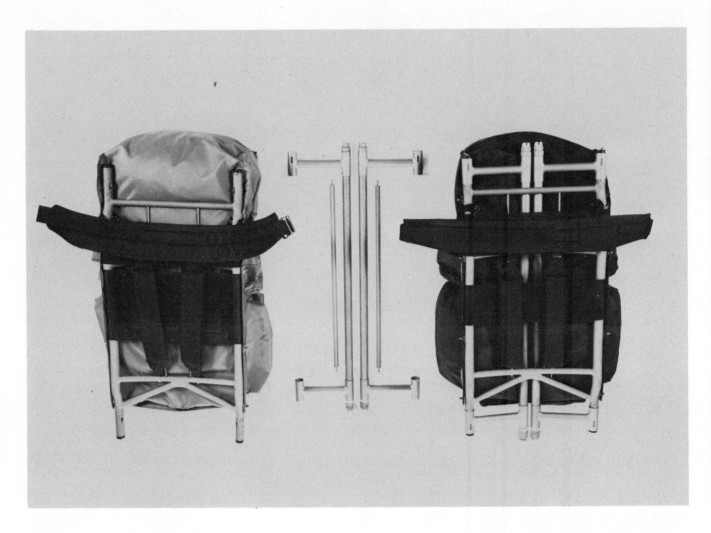

By making a cot frame a part of a back-pack frame, designers Keck-Craig Associates devised a way for a back-packer to carry a cot with him, one that can be set up easily anywhere, by someone wearing gloves, in the dark. The cot's supports elevate it 3 or 4" (7.6 or 10.2 cm) above wet ground or the knobby undulations of rocks. These supports are concentrated under the cot's midsection, so it needs only a short stretch of level ground for support.

To assemble the cot at a campsite, a hiker merely detaches the cot frame from the backpack frame by turning simple attachment clips; extra tubing slips from within the cot and pack frame tubes and is fitted together into a frame over which the cot's nylon body support fits. Cot frame and body support weigh only 2 pounds (.9 kilograms), and of course, should a back-packer not want to travel with the cot, he or she merely leaves it behind. Assembled the cot measures 27 x 75½" (68.6 x 191.8 cm) and the backpack frame 16¾ x 34" (42.5 x 86.4 cm).

Materials and Fabrication: cot and pack frames: 1.125 OD anodized aluminum tubing; backpack and cot body support: Para-Pac close woven nylon. Cot material has Velcro fastenings; frame ends are closed with molded acetal caps (injection molded) on stainless steel chains.

Client: Jeffrey F. Jagels, Pasadena, California.

Consultant Design: Keck-Craig Associates: Henry C. Keck, supervising partner; Thomas J. Campbell, chief industrial designer; Roy K. Fujitaki, Don F. Carter, product engineers.

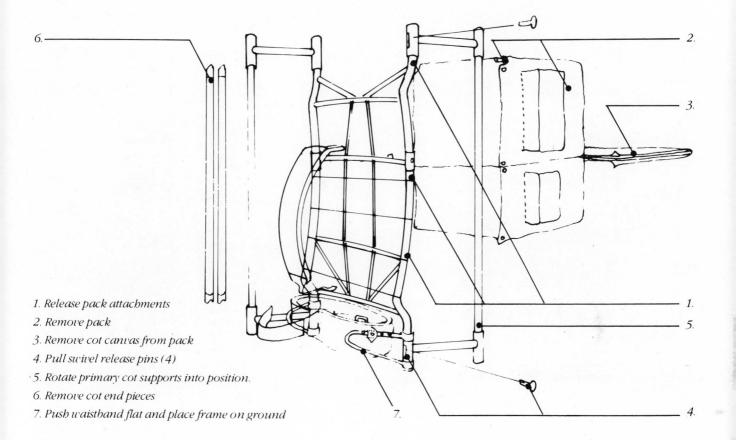

1. *Release pack attachments*
2. *Remove pack*
3. *Remove cot canvas from pack*
4. *Pull swivel release pins (4)*
5. *Rotate primary cot supports into position.*
6. *Remove cot end pieces*
7. *Push waistband flat and place frame on ground*

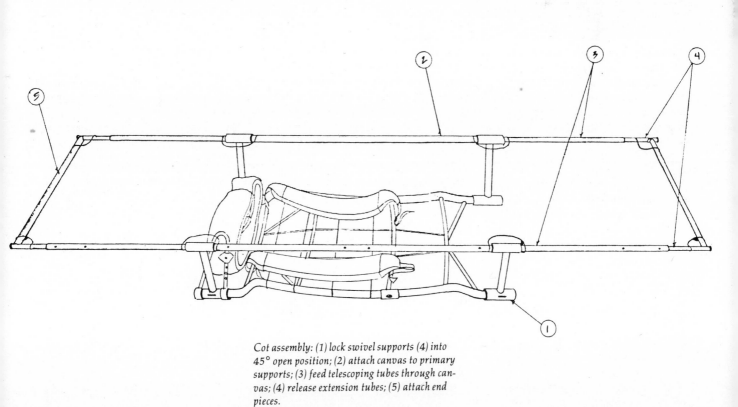

Cot assembly: (1) lock swivel supports (4) into 45° open position; (2) attach canvas to primary supports; (3) feed telescoping tubes through canvas; (4) release extension tubes; (5) attach end pieces.

CONTRACT AND
RESIDENTIAL

Howell Deschamps I Chair
Brickel 2412 Alexandria Chair
Brown Jordan ''Cricket'' Lounge Chair
Stow/Davis Paradigm Office Seating
Fiberesin Industries Fiberlife Line of Eight Furniture Components
Trakliting ''Geometric'' Lighting Fixture
Prototype Toilet

Robert Ian Blaich

Robert Blaich is vice-president of corporate design and communications for Herman Miller, Inc., in Zeeland, Michigan. He lectures frequently on design, is a past chairman of the Michigan State Arts Council, and a board member of the American Federation of the Arts. His degree, from Syracuse University, is in architecture and design.

Michael McCoy

With his wife Katherine, Michael McCoy is co-chairman of the design department at Cranbrook Academy of Art. He is also a partner in McCoy & McCoy, design consultants, whose current projects are the design of a Chrysler Corporation North American dealership prototype and a furniture system for Knoll International. McCoy lectures frequently on design and is a member of the editorial board of *Industrial Design* magazine.

Jeffery Osborne

Since 1967 Jeffery Osborne has worked for Knoll International, where he is currently vice-president of design. His control extends to all Knoll products, except textiles, in the U.S., Europe, and Japan. Osborne got the job offer from Knoll while he was studying marketing at Michigan State University, but he waited until completing his studies before joining the firm. The studies stretched out to include a fifth-year, following his BS, at Michigan State, during which he took senior design courses, and following that, study at Wayne State for a master's of business administration. In 1970 Osborne moved to Knoll's New York office as manager of marketing services. In 1976 he became Knoll's director of marketing for products, and in 1977 he was made vice-president.

Lela Vignelli

Born in Udine, Italy, Lela Vignelli received the Dottore Architetto degree from the University of Venice's School of Architecture and in 1958 studied architecture under a fellowship at the Massachusetts Institute of Technology. The following year she went to Chicago to work in Skidmore, Owings & Merrill's interiors department.

In 1960 she returned to Italy, opening with her husband, the Lela and Massimo Vignelli Office of Design and Architecture in Milan, where she did institutional and corporate work in interiors, furniture, product and exhibition design.

In 1962 she became a registered architect in Milan and in 1965 joined Unimark International Corp. in Milan, moving back to the U.S. to become executive designer for interior design in Unimark's New York office. Currently she is president of Vignelli Designs and vice-president of Vignelli Associates, New York.

The Vignellis received the 1973 Industrial Arts Medal of the AIA.

Introduction

Among the 35 entries (including 10 prototypes) in this year's Contract and Residential category, the jurors—Robert Blaich, Michael McCoy, Jefferey Osborne, and Lela Vignelli—found nothing new in technology or form. And though their initial reaction to this dearth was a mixture of annoyance and resignation that sent them paging through piles of 1978 magazines looking for innovative items they were sure had been designed in their field during the year, they were able to add only a few things, notably Ward Bennett's Alexandria chair and Trakliting's "Geometric" fixture, to what had already been submitted for review.

The jurors could only conclude, reluctantly, that contract and residential designers had spent the year refining rather than innovating. Eames, of course, always considered himself a refiner rather than an innovator, probably because, though his designs were wonderfully innovative, he spent the majority of his time refining what he had innovated.

But that 1978 was a year of refinement in the contract and residential fields seemed ironic, for the nature of the furniture business works against refinement. Aside from imitation, which, of course, is part of a refinement process, the push, the itch, the overwhelming wish is for innovation: "How many times is this chair going to be restudied?" one juror asked impatiently. Refinement is not normally a task manufacturers do well. One juror described what he considered a recognizable, if not typical, furniture manufacturing process: "They hit the market with a new product; it sells, but it has drawbacks. They have hysteria, and they correct the glaring deficiencies. What they don't do is stop and ask 'What can we do to make this 5 percent, 10 percent, or 15 percent better?' "

While the jurors were willing to blame the American corporate institution for stifling true innovation, they were divided on the effect the system has on a captive designer. "I think for the most part very creative people in an internal group are terribly inhibited by the process," said one juror. "If you are day in and out with the marketing, sales, and engineering people, you are going to find yourself compromised." Moreover, he went on, "The information reaching a group of internal designers is determined by the information flow of the company they work for. Their focus is their company's immediate problems. Product design becomes a function of a company's manufacturing capabilities, and a captive designer's chances of taking a problem, standing back, and looking at it in a fresh way are circumscribed." Another juror said flatly, "I don't think there is fertile ground for creative design inside most corporations." However, the corporate training ground can be good for a designer, provided he doesn't stay too long, maintained another juror. "There are, of course, a lot of designers who have worked inside corporations, then gone outside and done creative things. The corporate experience has been part of their growth."

A victim of the refinement versus innovation discussion was Owens-Corning Fiberglas bathtub (T-16-60). Designed by an outside consulting firm, Richardson/Smith Incorporated of Worthington, Ohio, the tub showed considerable design refinement. One end is angled, for instance, and the top edge faceted, so a bather can soak in a reclining position with both head and back supported comfortably. The designers suggest that the flat top tub rim helps bathers get a good grip when lifting themselves from the tub or lowering themselves into it. And they minimized decorative lines, eliminated ribs in the molded-in soap tray, and eschewed other detailing to make the tub easier to clean. Moreover, they widened the tub bottom so it can accommodate most everybody ("90th percentile of the hip population") and flattened and textured the bottom surface so someone standing on it is safer. Outer wall construction leaves toe space beneath the tub's outer edge so someone can stand close to the tub and bend over it more easily. Compression molded from sheet molding compound (inorganic-filled polyester and thermoplastic which is fibrous-glass reinforced) with a molded-in slip-resistant texture, the tub cuts costs to where, at a suggested retail

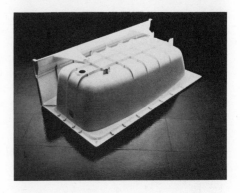

Owens-Corning Fiberglas T-16-60 Tub. Manufacturer: Owens-Corning Fiberglas Corp., Toledo, Ohio; staff design: Scott A. Calvert, Merritt W. Seymour; consultant design: Richardson/Smith Inc., Worthington, Ohio: Deane W. Richardson, Nilo-Rodis, David D. Tompkins.

price of $93, it is competitive with porcelain enameled steel tubs.

But the jury felt the manufacturer and designer had missed a chance to be truly innovative—to rethink bathtub design. "It is certainly not a breakthrough," mused one juror. "It's only value is that they did carefully consider some minor things that could be changed—the total interior space and the toe space beneath the exterior side. What I don't like are the crisp edges, part of the early 1960 American design vocabulary, which are just on their way out; they don't improve the grip as the designers explain." Said another juror, "My reason for voting no is that a large corporation had a chance to go the next step, to take their material and do other things with it, mold-in hand grips, for example, not only for elderly people but for young people too, for anybody who has to get into a machine like this. I find the tub's hard edge difficult to accept. It is merely a ring built around the bath-tub; that's all."

"There are opportunities for manufacturers," the jury stressed, "to go the next mile through research and development and not just satisfy the decisions of a perfunctory sales meeting."

Too often even when this year's contract and residential designs showed refinement, their refinement was limited. Often it touched only one particular problem in the range of problems posed by seating and lighting. It was as if, in many designs, the designers had chosen to address a single isolated problem, that of cost, say, or machining, or conformity to a host of body types. None of the entries, the jury felt, offered positive solutions to the whole spectrum of seating problems.

These entries, of course, came solely from American designers and corporations, and at least one juror felt that while he had seen nothing in the United States during the year advancing the state of the art, he had in Europe. In items introduced at Oratex, for instance, he had seen things, he said, that will set trends for a year or so. German manufacturers in particular, he felt, are making advances in office seating, because of government regulations. These regulations legislate style, or form, for economic reasons. The result is what one juror called "repeating form." He explained, "If you have a high-tech CRT unit, you must have a high-tech chair to go with it." Somewhat like the sport's shoe—which in the past few years has gone from an all-purpose shoe, the sneaker, to specialized shoes for running, hiking, boating, basketball, tennis, and, perhaps by now, sitting—the chair in Germany is undergoing design fragmentation. But the quality of the result is fragmented, too, because they "forget that the chair must move and support different-sized persons in different positions." Though the Germans design high-tech well, because they've been refining that particular problem for a long time, the jurors insisted, "We have to get more humanity back into the solutions. There should be more talk of wood finishes and stock and a divergence from machinelike problems. The enigma is that offices are getting more factorylike, more machinelike, and the products in them tend to go in that direction."

Perhaps most disturbing to American perception is that the Germans too often stress high-tech and economics at the expense of aesthetics. Though the jurors seemed unanimous in their determination to select only aesthetic forms for this review, they were equally adamant that a chair or other piece of furniture must ultimately be judged in its appropriate environment, performing its intended task.

Elsewhere in Europe aesthetics and function are a focus of design, the jurors went on. Europeans are generally ahead of Americans in considering function, they maintained. "Somebody who sits in front of a computer readout terminal all day has different seating needs from someone lounging at an airport, and in designing for these differences the Europeans, primarily the Italians, have taken the leadership." Unlike the Germans, the Italians and other Europeans are trying to design chairs that fit any environment. Instead of having separate chairs for separate functions, they seem to be saying, why can't you have a chair that suits a conference room, an executive office, a secretarial or a machine operation niche with equal ease, a chair that goes beyond current fads of fashion, scale, and color. "In the U.S. the furniture industry is years behind the possibilities because the furniture industry is taking a conventional approach to what furniture should be," said one juror.

What are some of the directions in which U.S. contract and residential furnishings design may move? If there was agreement on the future, it was that the office will probably continue as a space where people come together to communicate and that office hardware design would concentrate on equipment that will bring progress, make it possible to do things, such as write a memo or record a thought, much easier and faster.

Within this context furniture will become even more symbolic, though perhaps moving beyond its present symbology of status. A recent survey by Harry Paul found that office workers wanted more say about the office environment, that they wanted, for one thing, more privacy, a little space they can call their own. So desks, for instance, will be designed to provide a sense of place, of individual territory. "There are corporate people thinking of office space in fresh, bright ways," stated one juror, "ways that could help offices humanly reinforce the people working in them. The changes will probably come quickly."

An elegant, upholstered, stacking armchair, the Deschamps I is meant to fill a niche, however slight, in the contract furniture field. Most stacking chairs do not have arms. The Deschamps I does. It also has a combination of clean lines, chromed steel, and upholstery over a formed plywood shell that make it comfortable and elegant. The designers used three pieces of oval steel tubing plus a ⅝" (15.9 mm) spreader rod welded in six places and chromed to form a frame. Within this frame they sling a seat shell of ⅜" (9.5 mm), 5-ply, formed plywood, covered by foam and upholstery. The designers arched the tubing slightly along the chair's back to avoid a sagging look, which happens when the back line is straight. Dimensions are 30⅛ x 20⅜ x 25½" (76.4 x 51.7 x 64.8 cm). Seat height is 17½" (44.5 cm). Price: $162 to $176 depending on fabric.

Materials and Fabrication: frame: 11/16 x 1⅜ x 12" (1.7 x 3.5 x 30.5 cm) gauge steel tubing (electro welded, bent on Pines Bender). Seat and back: ⅜" (9.5 mm) 5-ply wood; padding 1" (2.5 cm) polyurethane; upholstery: customer's choice.

Manufacturer: Howell Co., a division of Burd, Inc., St. Charles, Illinois.

Staff Design: Richard Lyons, contract sales; John Gall, production engineer.

Consultant Design: Deschamps Mills Associates, Ltd.; Robert Louis Deschamps, president-director; Kenneth Schory, staff designer.

The Alexandria chair that Ward Bennett designed for Brickel Associates comes also as a sofa. As a sofa it has the same tight upholstery, the same white ash outline, and the same 25" (63.5 cm) back height. The sofa is wider, of course, 88" (223.5 cm) in all, wide enough to seat a pigmy band should one want to sit in the lobbies and reception areas the sofa is meant to enhance and serve. It is perhaps a mark of the design's success that the chair manages to look light and simple, while appearing definite, almost stolid and obtrusive. This near obtrusiveness is seemingly intended, allowing the chair to be a signal, announcing a *place* to sit as well as offering something inviting to sit upon. It does both with grace and aplomb.

Materials and Fabrication: solid hand-rubbed white ash components (carved, doweled, and glued).

Client: Brickel Associates, Inc., New York.

Consultant Design: Ward Bennett.

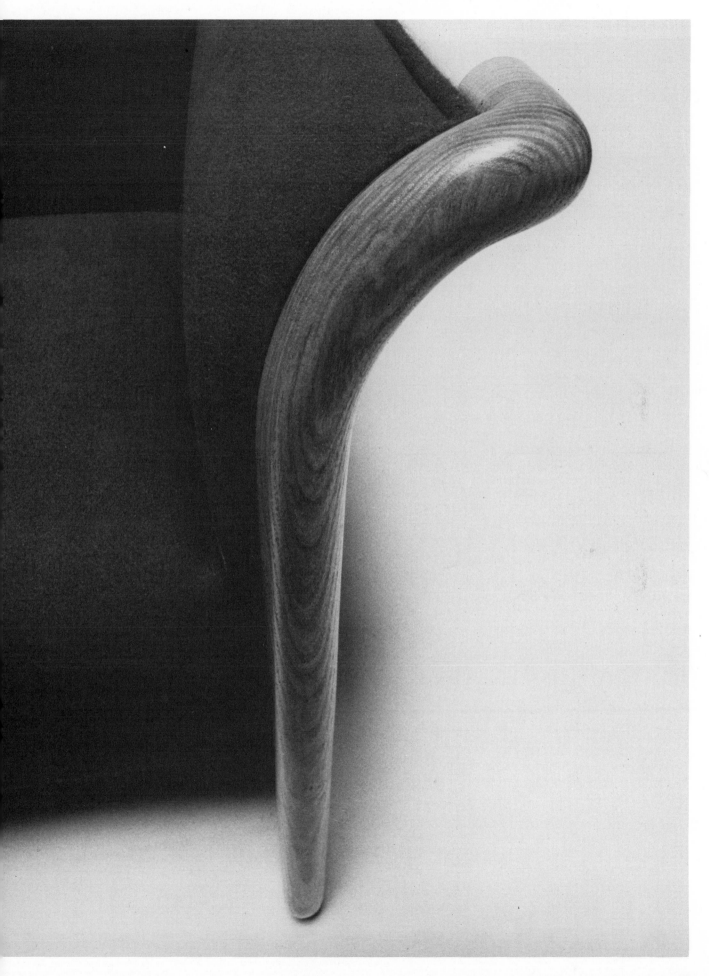

Meant to be used indoors or out, the Cricket Lounge Chair is designed to stand up to the elements and to fold flat, presenting a surface 27 x 23 x ¾" (68.6 x 58.4 x 1.9 cm) for easy storage in pool houses, boats, tool sheds, or already jammed closets. When open the Cricket's shape is what its designers call a "stressed polygon"—its tubular aluminum framing-spokes held in place under tension against one another by high tensile strength nylon straps. Vinyl-coated polyester mesh forms the sling seat. When open the Cricket's dimensions are 27 x 26 x 31½" (68.6 x 66 x 80 cm). The chair seat height is 15" (38 cm); seat depth is 19½" (49.6 cm); and height of the back is 19" (48.3 cm). Suggested retail price is $93.

The jury praised the Cricket chair's suspension, which lets it slide forward as you lower yourself into it and as you get up. Dissatisfaction came from the fact that the side straps, connecting the top of the back with the front edge of the seat, look like traditional, functional arms, when in fact they are not. And one juror pointed out: "You have a few hard points exactly where you should not have hard points." But he saw the chair as aesthetically appealing and as "having good material. It is not a new concept, but you see so many bad things coming out of that market that when something fairly presentable comes along you feel good about it."

Materials and Fabrication: frame: ⅜" (9.5 mm) aluminum rods and 0.83 gauge bend tubes of ¾" (19 mm) O.D. with an electrostatically applied baked-polyester finish. Seat sling: vinyl-coated polyester mesh. Straps: closely woven high tensile strength nylon. Glides: injection molded polyester (these double as strap fasteners). Rivets: hollow stainless steel. Heliarc welded aluminum joints.

Manufacturer: Brown Jordan, El Monte, California.

Staff Design: Donald B. Colby, manager design.

Consultant Design: Henry P. Glass Associates: Henry P. Glass, principal.

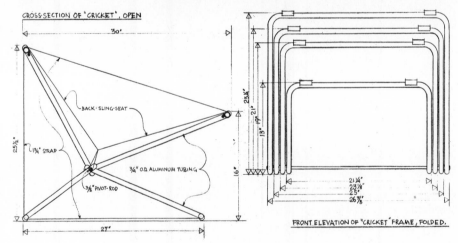

CROSS-SECTION OF "CRICKET", OPEN

FRONT ELEVATION OF "CRICKET" FRAME, FOLDED.

Stow/Davis calls its Paradigm chairs moderately sized. The smallest of them, found in the 170 series, is 22¼ x 27 x 34½" (56.5 x 68.6 x 87.7 cm) and is meant as a working chair for the compact work and conference spaces of open plan offices. At the top of the line, the 190 series is not much bigger, though it has padded arms, segmented upholstery, and an appropriately higher back. Measuring 25 x 28 x 54" (63.5 x 71.1 x 137.2 cm), it is meant for an executive office or boardroom. Chairs are, after all, still status symbols. In between is the 180 series, with padded arms and segmented upholstery but the same back height as the 170 series. The 180 is meant to seat managers. The Paradigm series has enough scope for an interior designer

to use it alone in corporate offices (especially open plan offices) where design continuity is important. And Stow/Davis plans to broaden the line by introducing, within a year or so, a version of the chair for secretaries and clerks.

Some of the Paradigm chairs have closed, upholstered arms, some open arms (both upholstered and non-upholstered), and one version comes armless. The jury was less enthusiastic about the armless version. Arms in this case lend the design substance and balance, they felt.

Designer Richard Schultz rounded all his Paradigm chair edges to protect clothes, limbs, and other furniture. And he softened the front edge of each seat to keep the chair from digging

into one's legs when it is tilted back.

Chairs can be vinyl, glove leather, or a range of lightly textured fabrics over foam rubber. Manufacturing is simplified by a design which eliminates all hand upholstery and almost all sewing. Retail prices range from $297 to $886.

Materials and Fabrication: seat, back, and side panels: structural high-impact polystyrene (molded), covered with polyfoam; seat and back cushions: high resilience polyfoam (molded). Arms: steel-tubing, mirror chromed or epoxy coated or with a rigid polyurethane, polyfoam wrap. Base: welded steel, chromed or epoxy coated.

Manufacturer: Stow/Davis Furniture Company, Grand Rapids, Michigan.

Consultant Design: Richard Schultz.

The Fiberlife furniture line features eight knock-down furniture pieces put together from particle board, wood dowels, canvas, and metal tubes. By standardizing fasteners and dimensions and using canvas where flat, durable surfaces are not essential, the designers reduced initial cost and maintenance. The line includes a dresser, a chest, a TV/stereo stand and end table, a record/cassette bin, a laundry hamper, a bookcase, and a three-bin catchall. Suggested list prices range from $29.95 for the record/cassette bin to $125 for the chest.

Though, as the jury pointed out, the technology for this type of furniture has been around for a long time and though their only comment on the design was that the combination of materials and the detailing was "okay," they were pleased that someone had brought this particular line to market, and they commended the price. "This kind of thing will break down people's expectation of what has to be," commented one.

Each unit is shipped knocked down, with assembly instructions and a tool. Buyers have a choice of canvas colors: yellow, rust, or brown.

Materials and Fabrication: Fiberesin ⅜" (9.5 mm) board, a high-density 55# particle board laminated on both sides with a melamine finish. Cotton canvas 10 oz/sq yd; steel tubing; wooden dowels.

Manufacturer: Fiberesin Industries, Inc., Oconomowoc, Wisconsin.

Consultant Design: Interface Design Group, Inc.; Leonard W. Kitts, president/design director; David L. Erickson, vice-president.

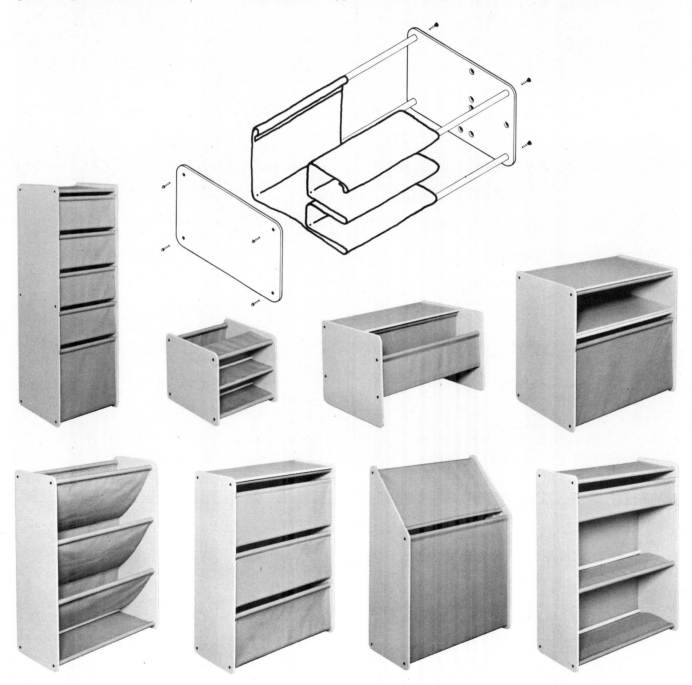

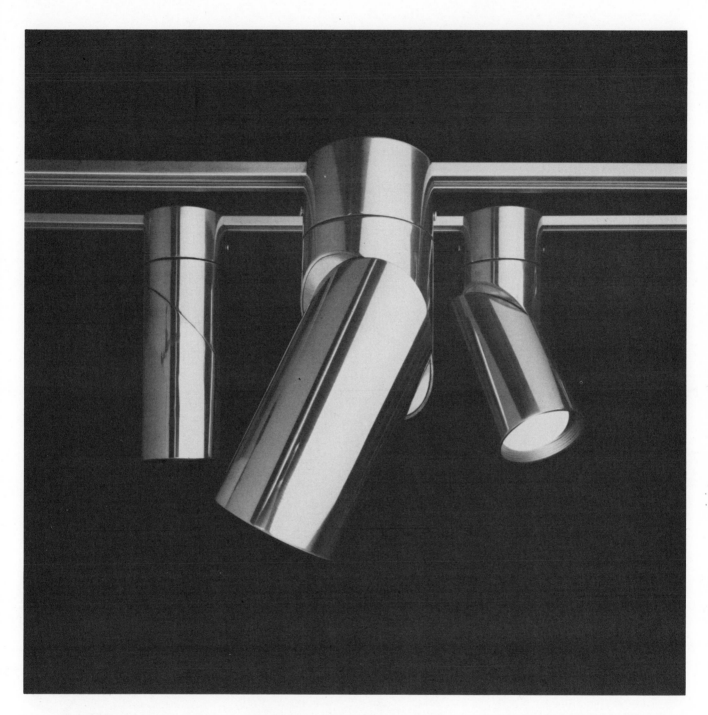

Consultant designer George Pelling conceived the Geometric track lighting fixture to get away from what he calls the traditional "theater" or "studio" look of track fixtures. He gave the Geometric a cylindrical aluminum body, carrying it past and around the ceiling track on which it's mounted. As a result the fixture is enclosed, not articulated as are its predecessors. Pelling's design carries track lighting through a familiar evolutionary leap.

So the light can spot a photo, a desk, a chair, a wall, or merely throw a pool of light onto the floor, Pelling lets the fixture rotate 360° on a horizontal axis and 360° along an axis penetrating the cylinder at a 45° angle. Moved about both these axes, the fixture will spot virtually anything around it. The Geometric comes in two sizes, one 11 x 3⅝" (27.9 x 9.2 cm); the other 13 x 4⅝" (33 x 11.8 cm). The larger unit retails for about $55, the smaller for about

$44. Pelling says the fixture conforms to all codes and listings—that safety was his primary goal.

Materials and Fabrication: tubes: aluminum (extruded); closures: aluminum (stamped); connector: Lexan 940. Aluminum is polished and brushed and the polished aluminum given a coat of clear lacquer.

Manufacturer: Trakliting Inc., City of Industry, California.

Consultant Design: George Pelling Design: George E. Pelling.

Developed by Notre Dame students during Armco's Student Design Program, in which students designed a total compressed air house, this prototype toilet flushes with compressed air and only 2 quarts (1.9 liters) of water, instead of the 6½ gallons (24 liters) needed by a more conventional toilet. Besides, these 2 quarts wash the toilet in a two-cycle rim flush, so the toilet needs no water-holding tank. That is one of the reasons for its appearance. The other is that the student designers left off a lid and a traditional seat, and by doing so made the toilet, they claim, more sanitary and easier to clean. The porcelain on which a user sits is more sanitary than the plastic or wood of conventional seats, they say; and by doing away with the seat attachments which catch dirt, they simplified cleaning. Toilet configuration is dictated by the support it intends to give a human body, allowing a user to lean forward comfortably. Dimensions: 18 x 28 x 16" (45.7 x 71.1 x 40.6 cm).

Materials and Fabrication: toilet: ceramic (casting) porcelain finish.

Consultant Design: Industrial Design Program, University of Notre Dame: Professor Frederick S. Beckman, project director; Joan Luttmer, student design group leader; Professor William Kremer, ceramist, technical assistant.

EQUIPMENT AND INSTRUMENTATION

Hewlett-Packard 300 Computer System
Hewlett-Packard 250 Business Computer System
IBM 8775 Display Terminal
IBM 3616 Passbook and Document Printer
Digital Equipment RT803 and RT805 Factory Data Collection
Terminals Workstations
MSA Lead-Foe II Air-Supplied Hood
Encon 160 Chemical Splash Goggle
Niranium Rollavue Sand Blaster
Fisher Accumet pH/Ion Meter, Model 750
GEX Intra-Oral Dental X-Ray
Datascope M/D3 Defibrillator System
American Red Cross Mobile Blood Collection System
Thermos Insulated Vaccine Carrier, Model 3500/8706
Medcor Lithicron-F Pacer Programmer
Picker Synerview 600 Computed Tomography System
John Deere 2600 and 2800 Semi-Integral Moldboard Plow

Dagmar Arnold

For 14 years Dagmar Arnold has worked for IBM as an advisory industrial designer in the Kingston, New York, office and holds several design patents and registered designs. She works currently in product design development, in design research and in planning and computer-aided design for display terminal development. She has also worked for General Motors and as a partner in a Swedish design consultant firm. After graduating with honors in industrial design from Pratt Institute. Arnold studied at the New Bauhaus School in Ulm, Germany, as a Fulbright scholar. She has taught at both Ohio State and Stanford universities.

Raymond Carter

Raymond Carter is design director of Charles Pelly Designworks in Los Angeles. He studied at Glendale College, Los Angeles Pierce College, and Art Center College of Design and has been a project director with Henry Dreyfuss Associates in both New York and Pasadena. In the past several years his design work has concentrated on medical instrumentation and transportation (including agricultural equipment) for U.S. and foreign clients.

Kenneth D. Collister

As chief industrial designer for the Ames Company, a division of Miles Laboratories, Inc., in Elkhart, Indiana, Kenneth Collister is responsible for the design of the company's complete line of medical laboratory instruments and their packaging. He is involved in each product from concept through manufacture, selecting materials, finishes, and methods of manufacture. Before joining Ames, Collister worked for Bell & Howell, the White Motor Company, the Richard Rush Studio in Chicago, and Mel Boldt & Associates where he designed radios and phonographs for Zenith. He holds several patents on motion picture and laboratory equipment.

James M. Conner
James Conner is one of four partners of Henry Dreyfuss Associates, where he has worked since 1952, first in California, then in New York. While with the Dreyfuss office he has spent more than two decades designing John Deere farm equipment, industrial machines, and consumer products and since 1960 has been responsible for the industrial design of all Polaroid cameras. Before joining Dreyfuss, Conner worked as an engineer for the National Advisory Committee for Aeronautics, NASA's predecessor. In the early 50s he took two years off to get a masters in design from Cranbrook Academy. He did his undergraduate work at Stanford.

Stephen MacDonald
Stephen MacDonald is currently account manager for Human Factors Industrial Design, Inc., New York. When he helped select the material for this review, he was managing editor of *Industrial Design* magazine. He entered the graduate industrial design program at Pratt Institute and took up a design career following 10 years with *The Wall Street Journal* as editor and writer.

Of all the categories, Equipment and Instrumentation was the hardest to judge. The five jurors—Dagmar Arnold, Raymond Carter, James Conner, Kenneth Collister, and Stephen MacDonald—spent a 12-hour day reviewing a bewilderment of computer systems, dental chairs, x-ray machines, food-serving carts, video display stations, data collection terminals, data encrypton devices, ultra-sound systems, defibrillation systems, tomography units, sand blasters, yarn tensiometers, friction meters, hydraulic rock drilling machines, typewriters, processors, printer-calculators, pressure sprayers, plastic-welders, plows, and particle flow monitors. It was not so much the number of entries (118) as the variety of fields represented—medical, office equipment, computer, agricultural, heavy machinery, etc.—and the need for the jurors to become instant experts in each that made the day long and the jurors indecisive. Time and again when asked to vote yes, no, or maybe on an entry, the jury's decision was a unanimous maybe. Somehow, 16 items were finally selected.

There was another reason for this unanimity. The level of equipment and instrument design is high enough that the elimination of one well-designed item and the selection of another is a matter not easily determined or defended. "We had trouble picking outstanding things," noted one juror. Part of the trouble was that the extraordinary was too often missing.

A plethora of good design is found in items such as typewriters and computers whose forms and appearances have been refined for years. "Computer boxes have been out a long time, and it's now really difficult to do one that's nicer than the rest," was a typical comment. A law of natural selection works just as surely with human-designed objects as it does with plants and animals. The computer that is successful in the market sets design standards for others, until eventually it is as easy to identify a computer as it is a tree or a fox. Still the jury felt that within a certain appearance group there is always room for refinement, for cautious if not sweeping design improvement, and where that ap-

peared, each juror tried to note it. Their search became one not so much for the extraordinary or the innovative as for the best of the good work submitted for their review. They did not feel, for instance, that either IBM's Electronic typewriter or Exxon's Intelligent typewriter showed such refinement. Though Exxon purposely made its typewriter look like one (it is reminiscent of the IBM Selectric), giving a typist a familiar form to work with, the jury felt the designers could have done better within this limitation. "I don't think they stepped up to any of the design challenges" was the general feeling expressed.

Despite the generally high level of equipment and instrumentation design, at least one juror was jolted by the cross-sectional glimpse of the field afforded by the submissions to this review. Viewing all the entries is "like looking at a slice in time," she said. "The things you've seen in the field or in other reviews that you've liked best stick in your mind, and you think that's what current design is," she went on. "But then you see a slice, and it's not all that rich."

Richness, like beauty, may be in the eye of the beholder, but whether or not American design has it, the jurors seemed in agreement that the climate for design present in Europe is missing here. Designers in Europe, the jurors felt, seem closer, more concerned and excited about each other's work. Moreover, the opportunities for designers to meet and discuss their work are more abundant in Europe. The chance to see the entire spectrum of design comes yearly in Europe at the Hanover Fair. By going to Hanover once a year "you can be up on everything that's happening. There's nothing comparable in the U.S." U.S. designers, the jurors felt, except those in major cities, work in greater isolation than their European colleagues.

What emerges from this review of equipment and instrumentation design is that its designers seem to be becoming more well rounded, gaining understanding, confidence, and strength in human factors, manufacturing, marketing, and engineering, but too often slipping in detailing. "They are doing better in organizing

components and working out configurations, making a lot of entries very good; but a sensitivity to integrating the materials and the form was missing," said one juror amidst general agreement. Detailing is difficult to control, they said, in partial defense of submissions whose detailing went awry. "I find in our office we sometimes loose control of the detailing at the end of a project. At the end the client is in a hurry. You rush and you loose details; the project doesn't come out quite the way you want it."

"It happens even if you think you're on top of a project," said another juror. "You think you have a design and you notice that a draftsman has arbitrarily changed an angle. You've got to fight it. And with smaller clients, you may turn over a finished set of drawings, sketches, and models, and when you see the manufactured product, it surprises the hell out of you."

It is easier to produce award-winning design by giving form to something that has lacked it before than it is by refining an already good item, the jury decided. An example of this is Gardner-Denver's Hydra Trac SCH3500, self-contained, hydraulic rock-drilling machine. But though the jury realized that the design by Tompkins & Associates of Boulder, Colorado, was a "tough job done against odds" (the odds are against industrial design permeating the rock drilling industry), and though the jurors wanted to commend the attempt, they felt they could not include an item that had gotten a sponge bath when it needed a shower.

More germane is the Fisher Accumet pH/ion meter (page 98). A redesign of a once blatantly undesigned machine, the meter came far enough in its design to gain the jurors' general approval, if not their overwhelming enthusiasm. It could, they felt, have come still further.

What other design developments became evident from this review? "Forms," the jury contended are becoming "very geometric with parallel sides and radius corners." At its worst the geometry is forced. But perhaps less aesthetically annoying but more trying for designers is that forms are becoming so simplified—boxes and circles predominate (though there is an incipient trend to larger curves)—and that the design depends on details. "They don't have a fancy form to carry the design along," one juror pointed out. "And the design depends on the graphics, switches, buttons, low gaps . . . that sort of thing."

This design simplification, with less reliance on frills or structural ornamentation, is, of course, seen in all facets of contemporary design. In trying to explain its prevalence in industrial design the jurors mentioned twin complexities: the complexity of today's life and the inner complexity of contemporary machines. "Life is complicated enough. We want simpler forms in our environment," offered one juror. If a designer makes a machine look simple it will seem less intimidating, becoming easier for an operator to approach and use.

But designers are, of course, more than form-givers. They are interdisciplinary dwellers who must speak the language of engineering, marketing, sales, administration, and materials handling. They must be able to see a machine in the context of its use, whether the machine is a tractor on a farm, a tomography system in a hospital, or a pH/ion meter in a laboratory. And they must be able to discover and understand that context. In addition a designer's understanding must remain supple. He is, for instance, being asked with increasing frequency to add to his lexicon that of the systems analyst. One juror told of designing a billing and record-keeping system for a hospital only to find there was no need for the system. The hospital director had merely thought there was.

Because of their knowledge and creativity designers will continue to come up with new product ideas. One of the entries the jurors studied, and ultimately rejected, was a computer scoreboard system designed for Colorado Time Systems by Aupperle Associates, Inc. of Loveland, Colorado. The designers figured out a way to segmentize each digit (to be flashed on the scoreboard) and arrange digits in modular grids, stacked in identical display lines so the board can be used for swimming, basketball, football, etc., giving it flexibility and, to a degree, portability. But the jury felt its aesthetics are not well resolved. It simply does not look as handsome as it might or should. A designer's chief business is still design.

Gardner-Denver's Hydra Trac SCH3500, Self-Contained, Hydraulic Rock-Drilling Machine. Manufacturer: Gardner-Denver Inc., Denver, Colorado; staff design: Jim Mayer, J.C. Pattel; consultant design: Tompkins & Associates, Boulder, Colorado: David Tompkins, Jack Hillman.

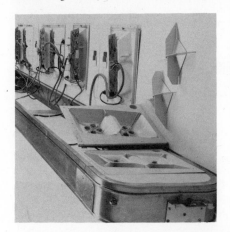 

System 300 Computer Scoreboard. Manufacturer: Colorado Time Systems, Inc., Loveland, Colorado; staff design: William T. Beierwaltes, James A. Hepp, Larry Bower; consultant design: Aupperle Associates, Inc., Loveland, Colorado: Donald P. Aupperle, Gary R. Paulson.

The way dishwashers and clothes dryers once moved into the average home, computers are now moving into the average business firm. Hewlett-Packard designed its model 300 with enough power and compactness to be taken out of the computer room and tucked next to a desk. But though the basic unit is compact (keyboard, print-out, screen, memory storage, and power supply), a business can add as many as 16 display stations and increase the removable disc storage more than 20 fold. Multiple printers also go with the HP 300. The *Design Review* jury liked the HP 300's clean lines, its uncluttered, simple look, and especially the detailing of its keyboard and readout area. "The details pop out at you," said one juror and another thought it a "perfect example of handling the details just right."

Dimensions are 43.5 x 24 x 33.5" (110.5 x 61 x 85.1 cm). Keyboard height is 28.1" (71.4 cm).

Materials and Fabrication: housing: formed sheet metal; display front panel and keyboard: injection-molded Lexan; paint: textured pearl gray.

Manufacturer: Hewlett-Packard Co., General Systems Division, Santa Clara, California.

Staff Design: David Horine, Scott Stillinger product design/engineering.

Consultant Design: Roger Lee/Design Associates: Roger Lee, president, design concept.

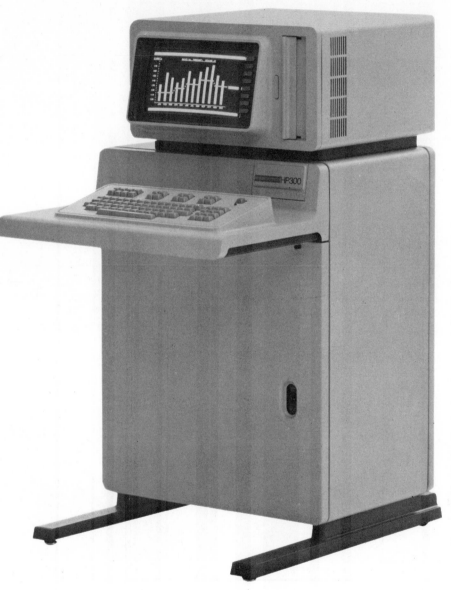

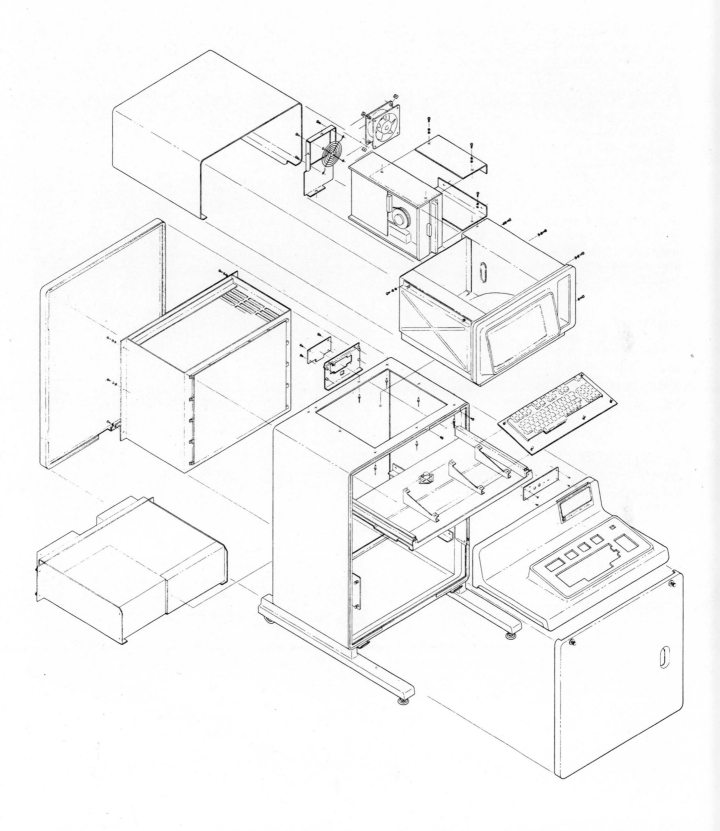

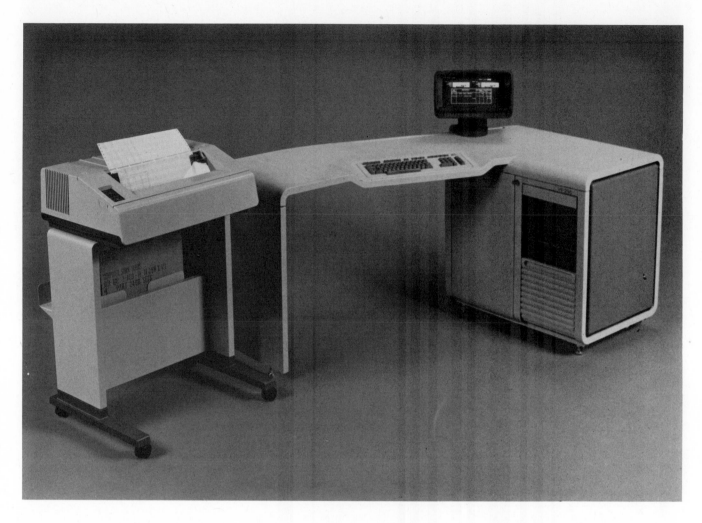

With a price of $24,500 the HP 250 is Hewlett-Packard's lowest priced computer. It is designed for use by departments within large companies and by relatively small businesses, which are, perhaps, computerizing their accounts for the first time. A typical operator is a clerical worker, not a programmer, using the machine for routine accounting work—order processing, payroll, inventory control, sales analysis, etc.—who sits at the machine for a full workday. It is meant to be used in an office, not a computer room, and for this reason the designers went to some lengths to reduce noise. It is cooled, for instance, by multiple fans, instead of one large, noisy one. In addition, the noise of the 250's printer can be muted by an optional enclosure. Its disc drives are kept clean by a tambour door. More than just a computer, the HP 250 is also a work station, and it is set up to be usable by anyone, regardless of their work habits. So oper-

ators can spread their worksheets wherever they want around them, the HP 250 has plenty of workspace on its L-shaped desktop, which stands 42" (106.7 cm) high, is some 21" (53.3 cm) wide and about 90" (228.6 cm) long. Moreover, the CRT display screen tilts, swivels, and moves in a track from side to side, so an operator can position it for optimum viewing. In contrast with the mobility of this screen, the keyboard is stationary, imbedded in the front edge of the desktop. While the jury thought the HP 250 more innovative than any other computer they reviewed, they felt, too, that the designers had been successful in making the machine appear simple and unintimidating. They liked the idea of a flexible screen and a fixed keyboard as well as the way the keyboard is integrated into the desktop. "It's a great solution," was an undisputed comment.

Materials and Fabrication: console top: sheet moulding compound; electronic enclosure and printer stand: fabricated steel; CRT enclosure, keyboard bezel, and tambour door: injection molded; accent panels: thermoformed.

Manufacturer: Hewlett-Packard Company, Fort Collins, Colorado.

Staff Design: Barry Mathis, industrial designer; Paul Febvre and Tom Bendon, product engineering.

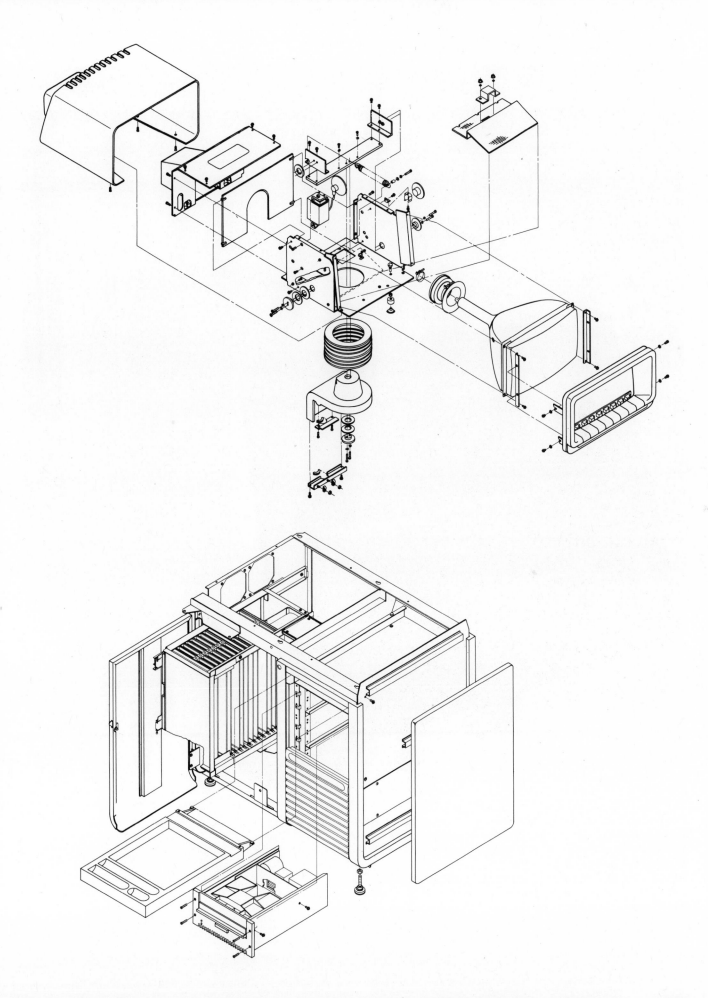

Meant for use as part of the IBM 8100 information system, the 8775 is a display unit on whose screen computer information appears. Its designers went to customers using other IBM computer display screens and as a result decided to make this one low enough so an operator can see over it. When it is used in a large office, the operator can still see and talk to operators nearby. Though low in profile, the screen is easy to see because it tilts up toward the operator, who can vary the angle of tilt, depending on his height and the height of the table the unit is on. To make the display easier to read, the designers gave the screen a low reflectivity coating. Unit's dimensions are 15 x 17 x 21" (38.1 x 43.2 x 53.3 cm). It weighs 46 pounds (21 kilograms) and sells for $2,835.

Materials and Fabrication: main enclosure: structural plastic foam (injection molded); display surround: homogeneous plastic (injection molded).

Manufacturer: IBM Corporation, Armonk, New York.

Staff Design: Edward Chamberlain, design manager; Mike Sharp, industrial designer.

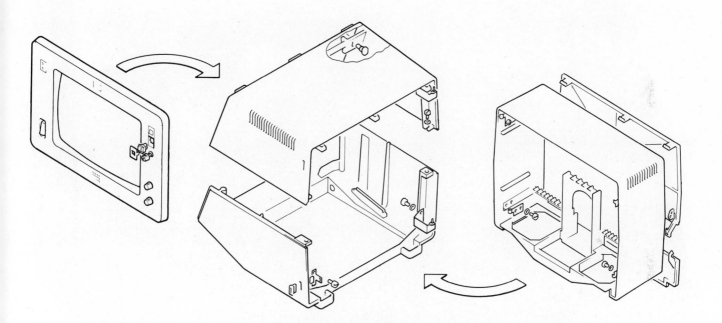

Used by savings and loan companies and banks, IBM's 3616 printer prints information on any size savings passbooks inserted in it. Or it will print on journal rolls or any of a host of cut form documents at a speed of up to 120 characters per second as part of the 3600 Finance Communication System. The 3616 is compact [20.9" (53.1 cm) wide, 14" (35.6 cm) deep, and sloping slightly from 10½" (26.7 cm) at the rear to 9¾" (24.8 cm) in front], and can be mounted either on top of a counter or sunk into it. This flexibility lets it be mounted so two tellers can use the machine, either sitting or standing. The designers also reorganized the controls, making them more convenient and more easily understood. They also made the print station easier to see and use. Retail price is $4,700.

Materials and Fabrication: top operator cover: formed sheet metal painted a fine textured black; front end cover: formed sheet metal painted a fine textured cloud white; within the top operator cover are: the operator control panel and passbook station, a journal cover, an access cover for ribbon loading, and a feature cover with extended document slot. These four parts are polycarbonate (injection molded) with an integral black satin finish.

Manufacturer: IBM Corporation, Armonk, New York.

Staff Design: Owen Shea, industrial designer.

Meant to be used in factories, Digital Equipment RT805 and its cousin (with slight control panel variations) RT803 data collection terminal workstations are rugged. They can withstand the heat, grease, grit, and dust of a factory and occasional rough use by whomever is punching information into them. The terminals are used to keep track of factory materials, inventories, and other shop floor data. They read badges and cards (up to 22 columns) and have a 32-character alphanumeric display, a 12-position key pad, and 5 function keys. The RT803 can double

as a time and attendance station.

If the designers wanted to protect the terminal from its surroundings, they also wanted to protect factory workers from the terminal. So they rounded its corners and recessed its control panel to keep someone who bumps into it from activating the keys, which are flush with the control panel surface. Units are gray with spatter-finished, black control panels that hide grease smudges. Dimensions are 17½ x 21½ x 12″ (44.5 x 54.6 x 30.5 cm). Units weigh 50 pounds (23 kilograms).

Materials and Fabrication: control panel: sand cast aluminum; sides and top wrap: castapoxy; backplate: 1/16″ (1.6 mm) brake-formed aluminum; bottom: ¼″ (6.4 mm) aluminum.

Manufacturer: Digital Equipment Corp., Industrial Products Group, Maynard, Massachusetts.

Staff Design: James S. Macconkey, industrial design coordinator; James A. Bleck, staff industrial designer; Ken Raina, engineering manager; Roger Gagne, product engineer; Ken Gulick, mechanical engineer.

Used for grinding lead or for sand or grit blasting, the MSA Lead Foe II air-supplied hood is an improvement over its predecessor, which was formed from aluminum. Now rotational molded of high-density, cross-linked polyethylene, the hood is dramatically better looking, and according to the manufacturer, protects a user better from falling objects while letting him see and breathe more easily. The designers used military human factors data in designing an internal suspension system that lets the hood fit most users comfortably, even those wearing glasses. If the face lens becomes scratched, it can be replaced without removing the hood. The hood is held in place by a soft nylon cape, which fits over the user's shoulders, extending to his belt. Hood dimensions are 10 x 14 x 11" (25.4 x 35.6 x 27.9 cm). Its corrugated vinyl hose is 36" (91.4 cm) long, and its suggested list price is $200.

Materials and Fabrication: hood: high-density, cross-linked polyethylene (rotational molded); visor lens: PVC (die cut); hose: corrugated vinyl (sewn); cape: nylon; valve (mechanical assembly).

Manufacturer: Mine Safety Appliances Co., Pittsburgh, Pennsylvania.

Staff Design: Elmer Buban, project manager; Frank Lotito, project engineer; Bill Phillips, design draftsman.

Consultant Design: Bally Design, Inc.; Alexander Bally, project leader; Tim Cunningham, designer.

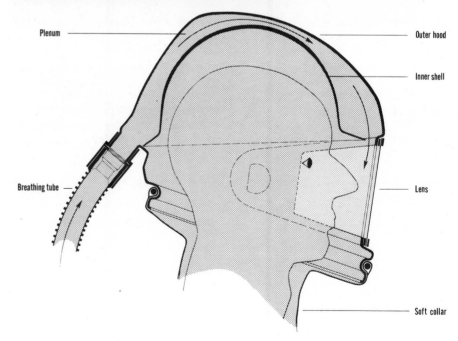

Plenum — Outer hood

Inner shell

Breathing tube

Lens

Soft collar

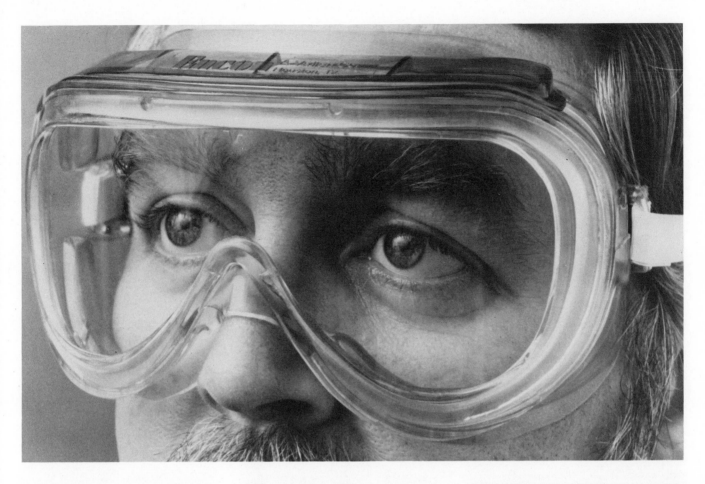

According to the manufacturer, the Encon 160 Chemical Splash Goggle lets someone wearing it see better and be better protected from splashing chemicals than does any other goggle. Not only does a wearer have 160° vision compared with an average 100° offered by other goggles but also the Encon 160 goggles can be worn with glasses or even outsized safety spectacles. At the same time their flexibility holds them snugly against most faces, regardless of that face's configuration. Moreover, the Encon 160 goggles are said to have 4 or 5 times the air openings of similar goggles, offering better air circulation around the face. Should chemicals find their way into the goggle, they are kept away from the user's eyes by a system of labyrinths and drains. Suggested distributor prices per goggle run from $1.90 to $3.30 depending on the type of lens used. Each goggle is 6½″ (16.5 cm) wide and 3½″ (8.9 cm) high. In preparing their goggles for a disparity of users the designers consulted anthropomorphic data from the U.S. Air Force, the Commerce Department, various university research papers, and ophthalmologists.

Materials and Fabrication: lens: .060 polycarbonate (injection molded); body: flexible vinyl (injection molded).

Manufacturer: Encon (A Vallen Company), Houston, Texas.

Staff Design: Robert Wright, manager.

Consultant Design: Goldsmith Yamasaki Specht Inc.: Paul B. Specht, executive vice-president, senior partner.

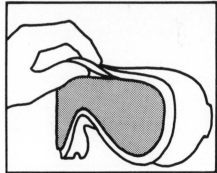

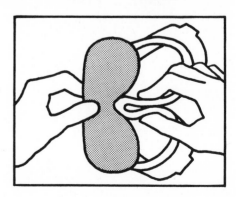

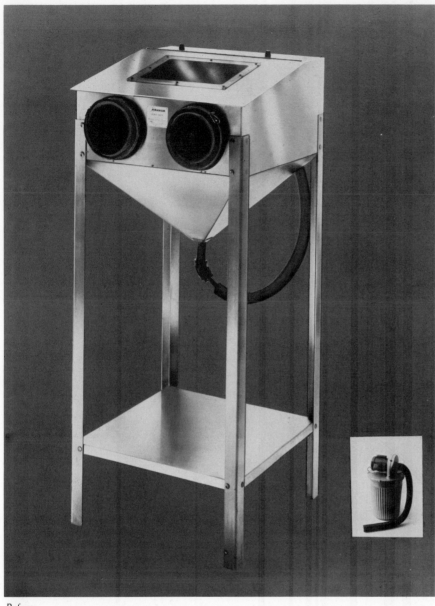

Before.

There's little question that redesign helped the appearance of Niranium's sand blaster, but the *Design Review* jurors thought the aesthetic improvement might have gone even further. They objected slightly to the blaster aping the appearance of a human face. At the same time they were impressed by how the design cut manufacturing costs and made the machine easier to use. Who wouldn't be? Not only does the machine cost 50 percent less to manufacture now but also its shipping weight is reduced [to 45 pounds (21 kilograms)], and it is tougher so, the Niranium Corp. claims, it suffers less damage in shipping. Besides all this the blaster has a disposable, plastic-film window. When the window becomes even slightly fogged from sand particles bouncing off it, the operator can roll a new soft window into place, tear off the old one, and throw it away. An operator slips his hands into the machine's yellow gloves and holds dental work beneath a nozzle. A foot pedal opens an air valve, letting sand blast through the nozzle onto the dental work. The blaster measures 56½″ (143.5 cm) high [43½″ (110.5 cm) to the glove pockets] and is 21½″ (54.6 cm) in diameter.

Materials and Fabrication: base: high-density HDL polyethylene (rotationally molded); cover, window frame, and glove pockets: ABS polymere (thermoformed).

Manufacturer: Niranium Corp., Long Island City, New York.

Staff Design: Douglass Mann.

Consultant Design: Carl Yurdin Industrial Design Inc.: Carl Yurdin.

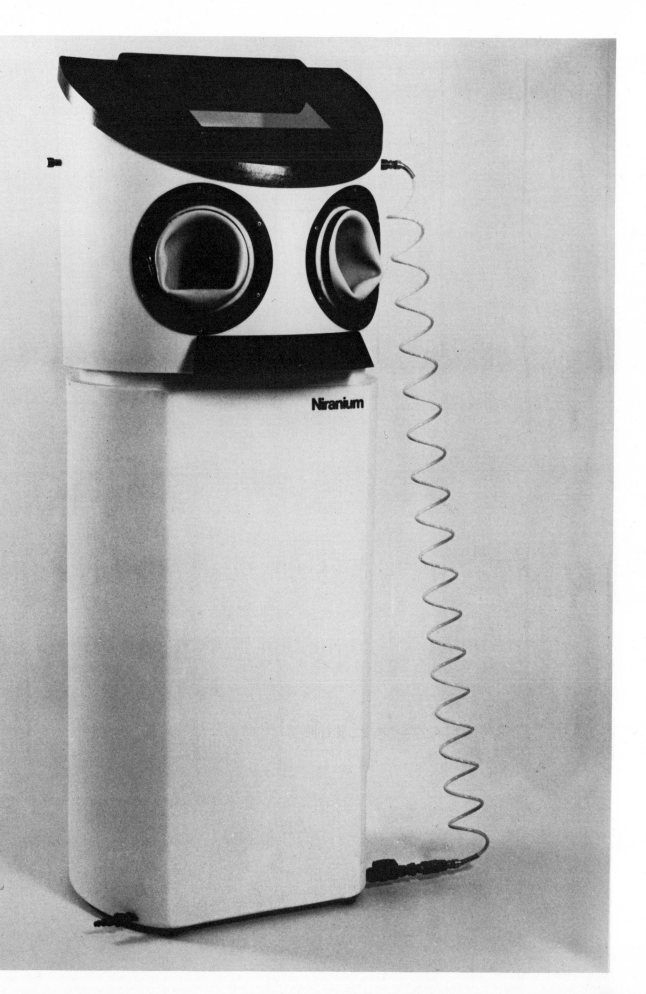

Fisher's Accumet 750 is equally at home in a research laboratory or a chemical plant. Fisher claims it will tell more about a chemical substance, with less fuss, quicker than other pH meters now on the market. Fisher's 750 adds microprocessing computation so the instrument presents an operator with a five-digit screen display of a solution's pH (acidity or alkalinity), adjusting the reading as frequently as 8 times a second. All this is accomplished at the push of a button. Besides the instrument will measure the chemical-electrical reaction within a solution in millivolts, give a continuous reading of its temperature (in centigrade or Fahrenheit), and measure its changing strength or concentration.

Fisher designers changed the originally intended die casting to vacuum-formed Rohm & Haas Kydex [.125" (3.2 mm) thick], cutting tooling and production costs, while meeting chemical resistance and flammability standards. Model 750 measures 11 x 13 x 4½" (27.9 x 33 x 11.5 cm); weighs 8 pounds (3.6 kilograms), and carries a suggested retail price of $1,550.

Materials and Fabrication: front panel: Duralar H (Mylar) laminated panel, .040" (1.2 mm) thick; case: Kydex, .125" (3.2 mm) thick (vacuum formed with die punched holes).

Manufacturer: Fisher Scientific Co., Pittsburgh, Pennsylvania.

Staff Design: Donald R. Graham, corporate industrial design; Robert Purdy, electronics engineer; Dr. Dennis Gibboney, research application chemist; John Schneider, electronics design manager.

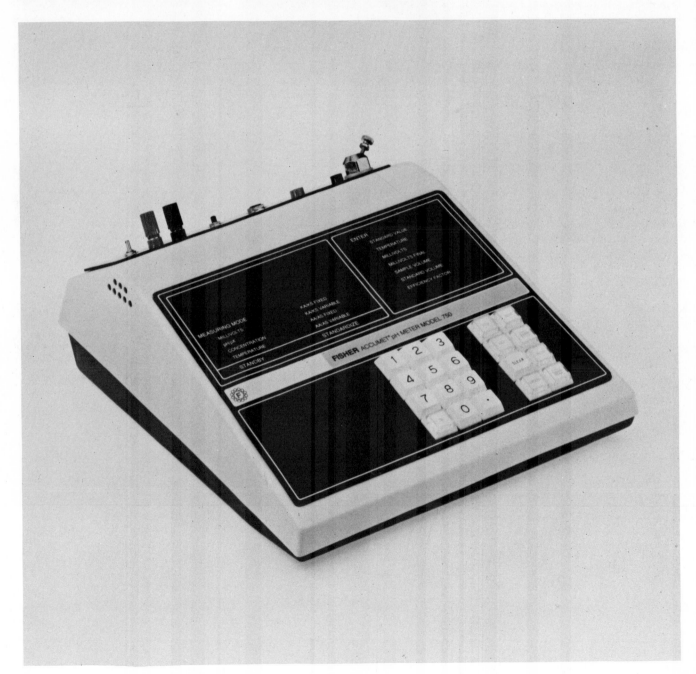

GEX takes x-rays of mouths and teeth in dentists' offices. Its designers meant it to be easy to assemble, clean, and use. The extension arm, for instance, has covered joints that won't pinch fingers as it extends and pivots, collapses and swivels to position the tubehead next to a patient's jaw. The tubehead is suspended from a short, one-sided yoke, making the tubehead easier to sight and position. The operator panel, which turns on the unit's power and sets exposure time, has a tough-activated surface, with no protruding knobs or buttons to hinder cleaning. When a technician wants to expose an x-ray, she merely depresses the hand switch at the end of a 6' (1.8 m) extension cord that lets her stand clear of the exposure zone.

Internally GEX's solid state electronics are modular. Eight circuit cards merely plug in, pulling out for service. The *Design Review* jurors liked the unit's simplicity and its consequently less threatening appearance, though they were not completely satisfied with the looks of the extension arm. "They've covered the springs and the joining mechanism, which is a nice way of cleaning that up," said one juror, "but covering the joints gives the arm a boxy look." And another juror pointed out that giving those joint covers a color different from the arms accented the joints, making them seem larger than they are.

Weighing 97 pounds (37 kilograms), the unit is relatively compact, the control panel is 28" (71.1 cm) wide and the suspension arm, when extended, is 69" (175.3 cm) long. The designers drew on anthropometric data found in Henry Dreyfuss's *Measure of Man* and N. Diffrient's *Humanscale 1 2 3*. Retail price: $4,935.00.

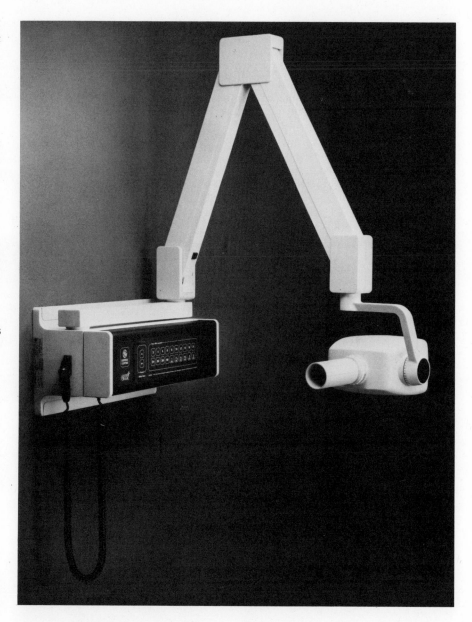

Materials and Fabrication: x-ray tubehead: aluminum permanent mold castings and plastic (injection molded); suspension arm: aluminum extrusion with sheet aluminum covers and GE Noryl structural foam covers over joints; control housing: aluminum permanent mold casting; control panel: suede Lexan touch button control panel 10 mm thick, adhesive mounted to sheet aluminum substrate.

Manufacturer: General Electric Co., Medical Systems Division, Milwaukee, Wisconsin.

Staff Design: Ed Roswog, industrial design project leader; Seth Banks, Jim Barnes, industrial designers.

Consultant Design: Goldsmith Yamasaki Specht Inc.: Paul Specht, executive vice-president.

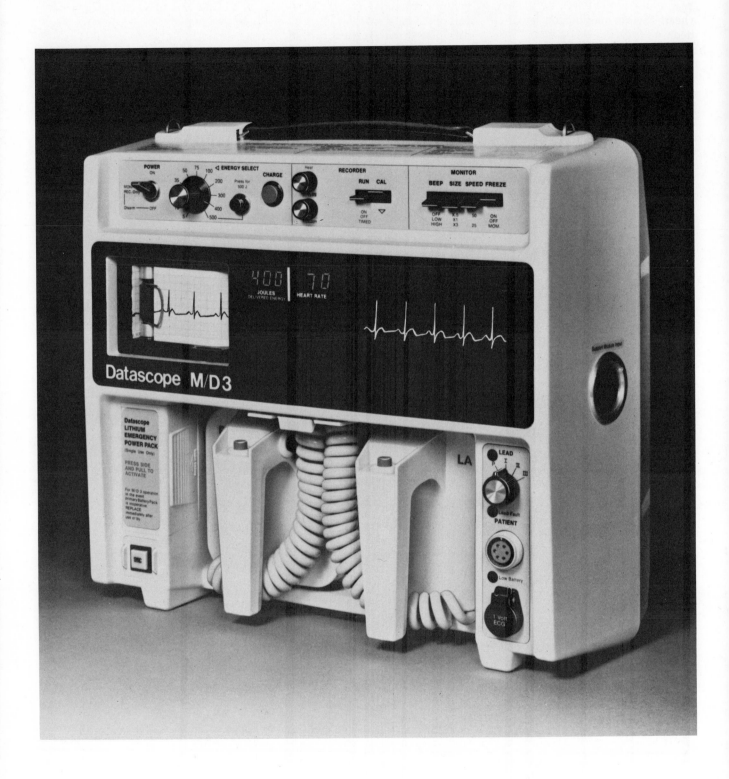

Carried by emergency medical units, defibrillation systems can identify, treat, and document heart fibrillation, a heart-rhythm malfunction associated with heart attack. Datascope's M/D3 defibrillator system weighs only 25 pounds (11.5 kilograms) and has, says it manufacturer, one major advantage over competing units: a lithium power source backing up its rechargeable nickel cadmium batteries. So the unit won't fade when in use. And it comes with a monitor that won't fade as the batteries discharge. If used in a hospital the unit can be plugged into a conventional electric socket. An operator merely turns the power switch on, then removes two hand-held paddles from the lower part of the system's case. These paddles, placed on the patient's chest, monitor his electrocardiogram (ECG). If the operator wants to record a portion of the ECG from the monitor, he merely pushes a switch, either on the paddle or on the control panel. To deliver a defibrillation shock the user positions the paddles, selects an energy level on the control panel, then pushes the paddle's discharge button. The portable system works in three positions: standing on its bottom edge, lying on its back, or propped at a 45° angle. Unit size is a compact 13 x 15 x 7½" (33 x 38.1 x 19.1 cm). Suggested list price is $5,995.

Materials and Fabrication: front and back covers: .125 polycarbonate (injection molded); control panel: polyester laminate; paddles: polycarbonate (injection molded); paddle handle: PVC finish.

Manufacturer: Datascope Corp., Paramus, New Jersey.

Staff Design: William Terrell, vice-president corporate design; David Schlesinger, industrial designer.

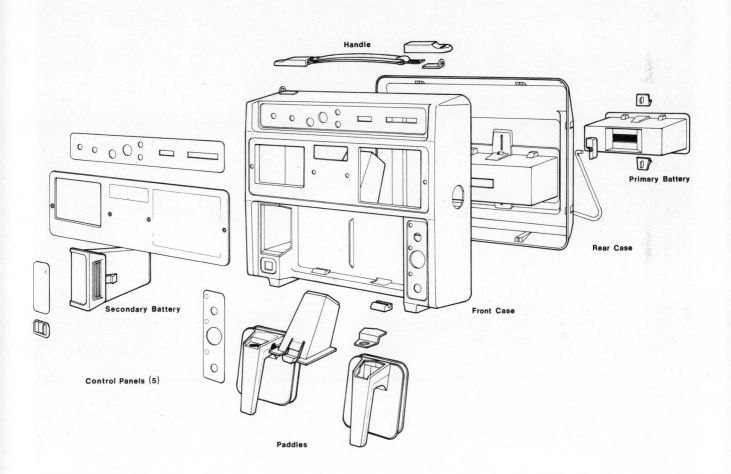

Handle

Primary Battery

Rear Case

Secondary Battery

Front Case

Control Panels (5)

Paddles

Each year the Red Cross gathers 5 million units of blood—about half this country's needs—and would like to gather more. Eighty-two percent of the Red Cross's blood comes from donors at temporary sites set up in offices, factories, churches, schools, and community centers. Each month Red Cross teams set up their equipment in 4,300 such sites throughout the U.S.

In 1970 the American National Red Cross Blood Program and the Rhode Island School of Design started looking at ways to improve the equipment carted from town to town by the Red Cross Bloodmobile. In use since the 1940s, this blood collection equipment—portable couches, tables, chests, and cartons—looked out of date, like World War I army equipment, and was hard to use. Some critics said the equipment discouraged donors, helping give them such a bad experience that only one in four gave blood again. The equipment, many thought, could be more comfortable, easier for the Red Cross staff to set up, take down, and use. The redesigned collection equipment, put together after eight years of research and design, went into nationwide service in March 1978. Adapting the Herman Miller Co/Struc cabinet system to bloodmobile use, the designers mounted some cabinets permanently on dollies, so they can be readily moved about. The cabinets, which must hold the 3,097 items used by the Red Cross teams at each donation site, all have modular, interchangeable drawers and trays, so drawers with fresh supplies can be slipped in without moving the whole unit. Moreover, cabinets, which have fold-up work surfaces on either side, double as desks and, when

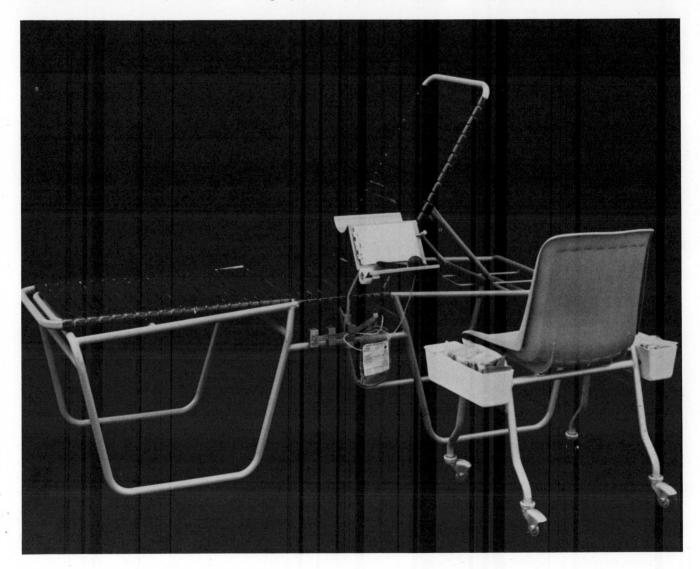

their lockable tambour doors are shut, as storage containers. In addition the designers developed a stackable chair on wheels. It moves on four casters, and when stacked, each stack can be wheeled about. The chair has a contoured polypropylene shell seat and tubular steel legs. Pods holding tourniquets and antiseptics, etc., used by nurses as they sit by a blood donor's side, attach to the frames on either side of these chairs. Donors lie on a lounge chair made of aluminum tubing and vinyl straps. This recliner adjusts at either end, supporting the donor in a semiupright position as he or she faces either end of the lounge. A nurse can then draw blood from either arm without having to move her chair to the lounge's opposite side. Some of these lounges have permanently attached wheels; others stack atop those with wheels to be wheeled on or off the truck. A typical bloodmobile carries $15,000 worth of equipment; 15 lounges, 1 recovery cot, 18 chairs, 3 work-supply cabinets, 24 portable blood refrigerators, 3 security containers, 1 handcart.

Materials and Fabrication: donor lounge: tubular aluminum with vinyl straps; wheeled nurses chair: tubular steel with polypropylene shell seat.

Client: American Red Cross Blood Services, National Headquarters, Washington, D.C.

Staff Design: Robert T. Schwartz, head, industrial design-architecture section; Leonard I. Friedman, head, biomedical engineering laboratory; Bernice I. Behnke, deputy director, blood services nursing.

Consultant Design: Rhode Island School of Design, department of industrial design; Marc S. Harrison, head of department; Edward Lawing, faculty advisor; 20 students, department of industrial design.

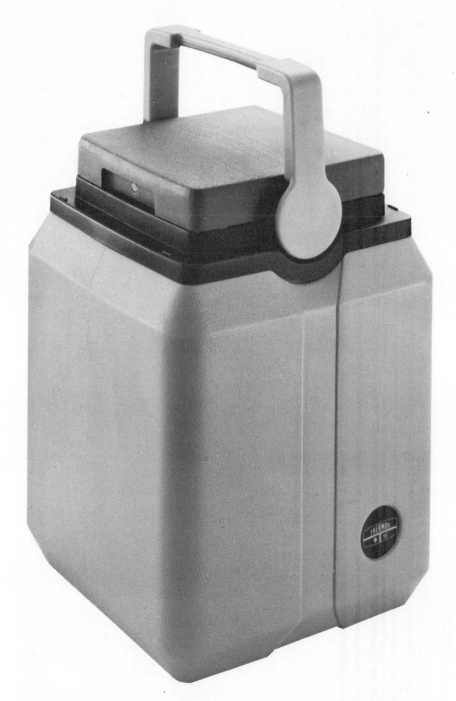

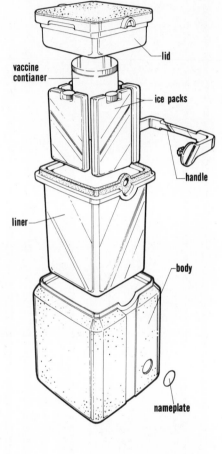

Insulated with polyurethane foam, the Thermos insulated vaccine carrier will keep biomedical products, such as vaccine or plasma, cold for as long as a day, according to the manufacturer, even in the tropics. Just as important, this polyethylene model is lighter and less expensive than its sheet metal predecessors. Four prefrozen ice packs fit into niches in the carrier's inner wall, surrounding the container with the vaccine or plasma to be kept chilled. The foam-filled lid locks in place when the handle is raised to its carrying position; with the handle down the units stack for storage—two features, borrowed from other Thermos products, which the jurors singled out for praise, though there was some grumbling about the way the center of the handle narrows. The jury liked the carrier's overall functional appearance. Without the ice packs in place the carrier will hold 4.35 liters (4.6 quarts). It measures 13'' (33 cm) high and 9½'' (24.2 cm) wide and weighs 4¼ pounds (2 kilograms), including ice packs. Sales price: $13.85.

Materials and Fabrication: outer body: .070'' (1.9 mm) polyethylene (blow molded); liner: .070'' polyethylene (injection molded); handle: foam-filled linear polyethylene (injection molded); lid: .070'' polyethylene (blow molded); ice packs: polypropylene (blow molded); name plate: screen-printed pressure-sensitive Mylar.

Manufacturer: King-Seeley Thermos Co., Thermos Division, Norwich, Connecticut.

Staff Design: Richard H. Seager, designer, institutional products; Richard A. Tarozzi and Michael E. Laude, designers, consumer products; Peter J. Hadfield, product engineer, institutional products.

Medcor Lithicron-F Pacer Programmer

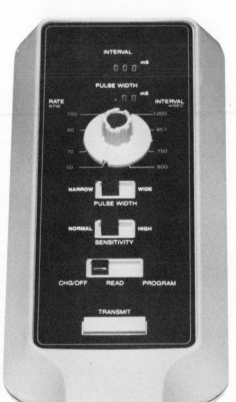

Heart surgeons use the Lithicron-F Pacer Programmer to adjust a patient's heartbeat once a pacemaker is inserted in his heart. On the programmer's display a doctor reads how the implanted pacer is functioning under different conditions: when the patient is under stress, say, or when he's at rest. If it needs adjustment, the doctor sets up a program on the programmer's control panel, turns the switch to "program," and transmits the new program to the patient's pacemaker. A hand-held instrument, the programmer weighs only 2½ pounds (1.3 kilograms) and is 12⅝" (32.1 cm) long and 3½" (8.9 cm) wide. The designers reduced the unit's weight, making it, they claim, 60 percent lighter than competitive items.

Materials and Fabrication: housing: Hercules grade 6532 Type H polypropylene (injection molded); control panel: die cut .030" (0.8 mm) Panvin with a black Panel-Tex finish (screen printed) [panel is supplied by Panelgraphic Corp.].

Manufacturer: Medcor Inc., Hollywood, Florida.

Staff Design: Robert J. Drozdowski, manager, medical product development, ESB Technology Center; Peter Schuman, Sue Scharples.

Consultant Design: Robert Hain Associates, Inc.: Peter F. Connolly, vice-president, packaging coordinator and designer; Frederick B. Hadtke, senior vice-president, designer; Robert W. Hain, president, designer.

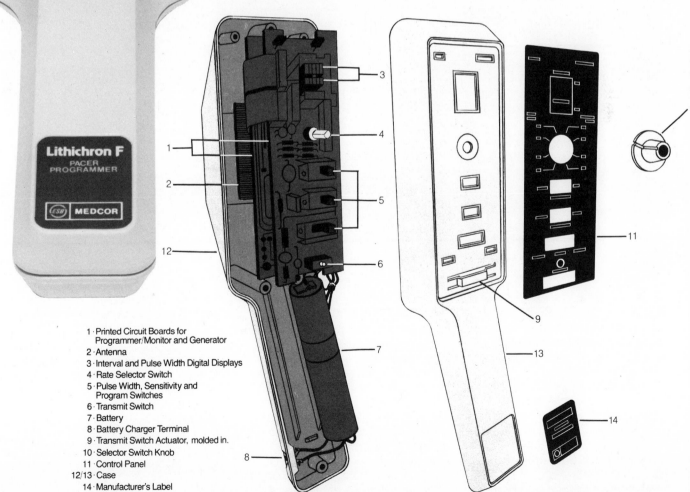

1 · Printed Circuit Boards for Programmer/Monitor and Generator
2 · Antenna
3 · Interval and Pulse Width Digital Displays
4 · Rate Selector Switch
5 · Pulse Width, Sensitivity and Program Switches
6 · Transmit Switch
7 · Battery
8 · Battery Charger Terminal
9 · Transmit Switch Actuator, molded in.
10 · Selector Switch Knob
11 · Control Panel
12/13 · Case
14 · Manufacturer's Label

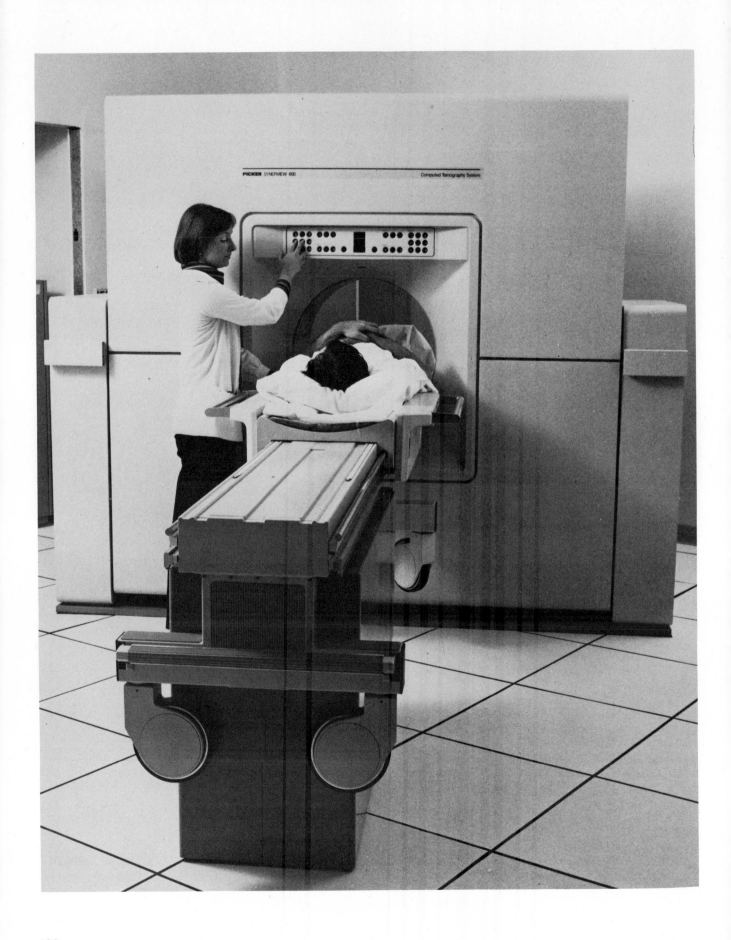

In one second the Synerview 600 produces a cross-sectional image of a human body. This image can appear variously on a CRT screen, in a color or black-and-white Polaroid print, or on an x-ray film with an accompanying computer analysis. And it can produce this cross section on a diagonal, tilting to take the scan while the patient remains horizontal.

Besides being faster than its predecessors, the Synerview 600 is easier to use. A redesigned control panel with dual controls lets an operator stand on either side of a patient while controlling the machine. Maintenance is made easier by hinging the entire front and back panels at the top, so they open upwards, floating above the gantrylike gull wings and revealing the machine's entire inner mechanism. And patient handling is improved, too. A roll-away cart collects the patient from his hospital bed. Each cart, designed as part of the system, is the exact height of hospital beds, and as a result, patients can be moved from bed to cart with minimum fuss. Armrests on the cart move to let patients on and off. Once on the cart a patient is pushed through the hospital to the scanning room and positioned for scanning by the tomography unit. Here the cart straddles a hoist which raises it to the scanning level and extends the cart, cantilevering it into the machine. The main unit has ribbed end panels with bumper surfaces to protect the gantry from bumps by carts moving through the scanning room. The gantry is 102 x 35 x 80¾" (259.1 x 88.9 x 205.1 cm). Suggested retail price: $750,000.

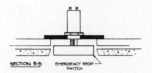

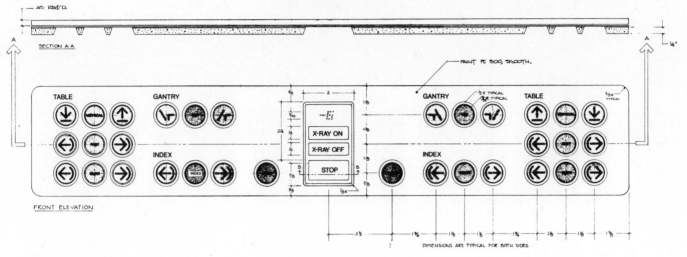

SECTION B-B   EMERGENCY STOP SWITCH

- AS REQ'D

SECTION A-A

PAINT PC 300, SMOOTH.

| TABLE | GANTRY |
| INDEX |

X-RAY ON
X-RAY OFF
STOP

| GANTRY | TABLE |
| INDEX |

FRONT ELEVATION

DIMENSIONS ARE TYPICAL FOR BOTH SIDES

NOTES:

1. CONTROL PANEL TO BE FABRICATED IN ALUMINUM, ACRYLIC OR SIMILAR 1/4" TH. MATERIAL. PAINTED FINISH.

2. TYPOGRAPHY FOR MAIN CONTROL GROUPS IS TO BE 16 PT. HELVETICA - UPPER CASE. SILKSCREEN ON PAINTED PANEL. COLOR - SEMI-GLOSS BLACK.

3. SYMBOLS USED ON TOUCH CONTROLS WILL REQUIRE CAMERA READY ART FOR PANEL MANUFACTURER.

4. SYMBOLS ARE TO BE REVERSE PRINTED IN PRODUCTION. THEY ARE TO APPEAR AS WHITE ON A BLACK FIELD.

5. TYPOGRAPHY ON TOUCH SWITCHES IS TO BE 10 PT. HELVETICA - UPPER CASE.

6. TYPOGRAPHY IN STATUS CONTROL AREA TO BE AS FOLLOWS: 'X-RAY ON' - 16 PT. HELVETICA - UPPER CASE IS TO APPEAR AS WHITE TYPE ON A RED FIELD. 'X-RAY OFF' 16 PT HELVETICA - UPPER CASE IS TO APPEAR AS WHITE TYPE ON A GREEN FIELD.

7. EMERGENCY STOP CONTROL - MECHANICAL SWITCH AS PER CALIF BENDER CORP. RED CAP WITH WHITE NOMENCLATURE 'STOP' IN 16 PT. HELVETICA, UPPER CASE, SILKSCREEN WITH PROTECTIVE WEAR INK. SWITCH IS TO LIGHT WHEN ACTIVATED. SWITCH IS MOUNTED THRU DEADFRONT PANEL.

8. RADIUS AT CORNER POINTS OF CONTROL PANEL TO BE DETERMINED BY PICKER IN CONJUNCTION WITH REQUIREMENTS OF FIBERGLASS OPENING PANEL. A 1/16" REVEAL AROUND CONTROL PANEL IS DESIRABLE.

9. CONTROL PANEL FINISH IS TO BE SMOOTH AND PAINTED TO MATCH GANTRY OPENING. PC 300.

10. A SLIGHT BREAK SHOULD APPEAR ON ALL EDGES WITH SPECIAL EMPHASIS ON OUTER CONTROL WELL EDGES.

11. ALL 'SYMBOL' CONTROLS ARE TO LIGHT WHEN CONTROL LIMITS ARE REACHED. LIGHT SHALL APPEAR AS RED THROUGH DEADFRONT APPEARANCE OF SWITCH. WINDOW FOR LIGHT WILL COMPLETE A CIRCLE AT OUTSIDE EDGES OF SWITCH. THICKNESS OF THE CIRCLE LINE TO BE 3/32".

12. PAINT CONTROL PANEL COVER. PC 300

13. REVIEW MODEL BEFORE FINALIZING PRODUCTION DRAWINGS. STOP BUTTON DETAILING HAS CHANGED AND THIS DRAWING SUPERCEDES MODEL FOR THAT CHANGE.

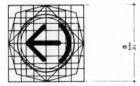

SYMBOL GRID CONSTRUCTION RELATIONSHIP SEE PICKER CORPORATE DESIGN SYSTEM GUIDELINES. GRID LAYOUT CONFORMS TO INTERNATIONAL MEDICAL EQUIPMENT COMMISSION GUIDELINES.

Control Panel

Materials and Fabrication: gantry: fiberglass lay-up; side pillars: cast aluminum with vacuum-formed, high-impact styrene end covers; gantry side covers: 16 ga sheet steel; control panel: machined acrylic; base 1'' (2.5 cm) steel plate. Patient-handling system: adjustable base covers: 16 ga steel (welded); outrigger legs: steel tubing; patient couch: lay-up carbon graphite; armrests: cast aluminum (machined); bumpers: molded rubber.

Manufacturer: Picker Corporation, Cleveland, Ohio.

Staff Design: Joe Stickney, general manager; Carl Brunnett, manager of engineering; Tony Zupanick, mechanical engineering manager.

Consultant Design: Richardson/Smith: Ed Lawing, project director, designer; Deane Richardson, chairman, designer; Keith Kresge, associate, designer; Larry Barbera, designer; Dave Tompkins, vice-president, designer.

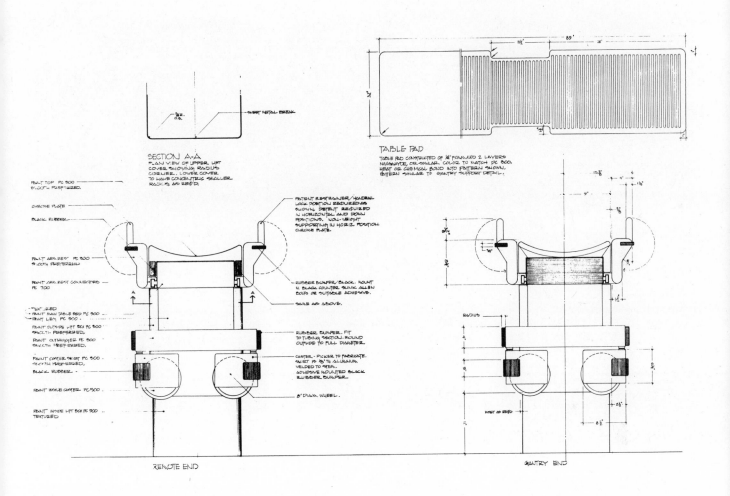

SECTION A-A
PLAN VIEW OF UPPER LIFT
COVER SHOWING RADIUS
CORNER. LOWER COVER
TO HAVE CONCENTRIC SMALLER
RADIUS AS REQ'D.

SHEET METAL BREAK

TABLE PAD
TABLE PAD CONSTRUCTED OF 1/4" FOAM AND 2 LAYERS
NAUGAHYDE, OR SIMILAR. COLOR TO MATCH PC 300.
HEAT OR CHEMICAL BOND INTO PATTERN SHOWN.
PATTERN SIMILAR TO GANTRY SUPPORT DETAIL.

PAINT TOP PC 300
SMOOTH PREFERRED

CHROME PLATE

BLACK RUBBER

PAINT ARM REST PC 300
SMOOTH PREFERRED

PAINT ARM REST CONNECTORS
PC 700

"TEXT" BED
PAINT MAIN TABLE BED PC 300
PAINT LENS PC 300

PAINT OUTSIDE LIFT BOX PC 300
SMOOTH PREFERRED

PAINT OUTRIGGER PC 300
SMOOTH PREFERRED

PAINT CENTER SKIRT PC 300
SMOOTH PREFERRED

BLACK RUBBER

PAINT INSIDE CENTER PC 300

PAINT INSIDE LIFT BOX PC 300
TEXTURED

PATIENT RESTRAINER/HANDRAIL
LOCK POSITION REQUIRED AS
SHOWN. DETENT REQUIRED
IN HORIZONTAL AND DOWN
POSITIONS. NON-WEIGHT
SUPPORTING IN HORIZ. POSITION
CHROME PLATE.

RUBBER BUMPER/BLOCK. MOUNT
W. BLACK COUNTER SUNK ALLEN
BOLTS OR OUTSIDE ADHESIVE.

SAME AS ABOVE.

RUBBER BUMPER. FIT
TO TUBING SECTION. ROUND
OUTSIDE TO FULL DIAMETER

CASTER - PICKER TO FABRICATE.
SKIRT 16 1/8" TH ALUMINUM
WELDED TO STEM.
ADHESIVE MOUNTED BLACK
RUBBER BUMPER.

8" DIAM. WHEEL.

RADIUS

INSET AS RELIEF

REMOTE END

GANTRY END

109

The moldboard plow is pulled by a tractor for primary tilling, turning up the earth before it is broken into smaller particles for planting. To do this heavy work the plow depends on six plowshares, which bite into and turn the soil. The designers' approach to this totally functional tool was to accentuate its form, to heighten the rhythm of its plowshares lined up along the main frame. They achieved this accentuation by painting the plowshares black against the frame's green. Further they tried to express strength and reliability by using rectangular steel tubing for the main frame and by giving simple, clean forms to all cast or forged parts. Then, to put the plow in tune with the earth it turns, the designers curved each standard (the metal arms holding the plowshares). Earlier designs had straight vertical standards, which the designers maintain "appeared at odds with the smooth curving path of the soil."

Materials and Fabrication: main structural components are of tubular steel; standards and coulters have cast components.

Manufacturer: John Deere Plow/Planter Works, Moline, Illinois.

Staff Design: W.W. Jackson, manager, planting, cultivation, and tillage, product planning worldwide; George Oelschlaeger, product engineer, supervisor; L.G. Arnold, product engineer.

Consultant Design: Henry Dreyfuss Associates: William F.H. Purcell, partner.

VISUAL COMMUNICATIONS

Dansk Teakwood Products Packaging
Copco Cutting Knives Sets Packaging
Myojo Foods "O My Goodness" Instant Oriental Noodle Packages
Myojo Foods "O My Goodness" Instant Oriental Noodle Marking
Dispenser Cartons
Pillsbury Panshakes Pancake Mix Package
Royal Crown Cola's KICK Packaging
J.A. Wright 7-lb Silver Polish Can Graphics
Rhodes "Beaver" and "Sunray" Steel Wool Packages
Gillette Max 1000 Hairdryer Package
McGraw-Edison Products Packaging System
GAF Corporation GAFMED Brand Identification and Packaging
J C Penney's Automotive Battery Line Graphics
J C Penney's Packaging for Assorted Kitchen Items
J C Penney's Leotard and Tights Packaging
J C Penney's Flasher Tags for Boys' Shorts
J C Penney's Merchandise Hang Tags for Children's Coordinated
Clothing
Container Corporation's Shape Op® Container
Container Corporation's Plastic Tank Sales Literature
Container Corporation's 1978 Promotional Calendar
Raychem Corporation Poster Series
IBM Corporate Recruitment Literature
Exhibition Catalog for the AIGA "What's Real in Packaging" Show
Girl Scouts Identification Program
National Shirt Shops Corporate Identity Program
CBS: On the Air Trademark
Raychem Corporate Poster Series
Three ICOGRADA Case Study Posters
Cranbrook Graduate Design Poster
Cranbrook New York Trip Poster
"Bass on Titles" Poster
Halleluiah Poster
Cluett, Peabody & Co., Inc. 1977 Annual Report
First Boston 1977 Annual Report
Potlach 1977 Annual Report
Hammermill Paper 1977 Annual Report
Filmways, Inc. 1978 Annual Report

Richard Danne
Since 1973 Richard Danne has been a partner in Danne & Blackburn, Inc., in New York City, and he is currently president of the American Institute of Graphic Arts. He has been an instructor of graphic design in the School of Visual Arts in New York City and an independent designer and consultant in Los Angeles, Dallas, and New York. Danne has a degree in fine arts from Oklahoma State University and attended the UCLA Graduate School of Design.

George T. Finley
Editor-in-chief of *Industrial Design* magazine for the past four years, George Finley is a member of the board of directors of Design Publications, Inc., the magazine's parent company. Since 1969 Finley has written about design and related subjects including product design, urban planning and transportation, packaging, graphics, and exhibits. Under his direction the magazine's approach to design has become more business oriented, relating design to the needs of corporations and at the same time giving human factors considerable attention. Finley has participated in each year's selection of design work for the annual *Design Review* book published under the auspices of the magazine. He has lectured at numerous schools, colleges, institutions, and design conferences and is a member of the editorial committee of the American Business Press and the advisory board of the Design Management Institute of the Massachusetts College of Art.

Gary Hanlon
Since 1976 Gary Hanlon has run his own graphic design office, The Design Center, in Boulder, Colorado. He grew to like Colorado while teaching for two years in the University of Colorado's Environmental Design College and returned to Colorado in 1976 from Ohio, where he had been a vice-president of Richardson/Smith, Inc., in Columbus. Originally from Cincinnati, Hanlon attended the University of Cincinnati's five-year design program, worked for Unimark International in Cleveland and Chicago, and was in charge of corporate identity work as design director with Saul Bass & Associates in Los Angeles before going to Colorado to teach. His work in corporate identity, packaging, and annual reports has won numerous awards, and he has spoken widely about graphic design at Canadian and U.S. universities.

Herbert M. Meyers
A specialist in marketing, packaging, and corporate design, Herbert Meyers is a principal in Gerstman + Meyers in New York City. He has done packaging and corporate identity programs for Ralston Purina, Black & Decker, H.J. Heinz, and Seven-Up, among others. He lectures and writes frequently about corporate identity and packaging design and is a board member of the Package Designers Council.

Patrick Whitney
Patrick Whitney is a designer, educator, and administrator, and he currently holds three positions that let him indulge each proclivity. He is a designer with the RVI Corporation, the executive director of the Design Foundation in Chicago, and an assistant professor at Illinois Institute of Technology's Institute of Design. Whitney's master of fine arts degree is from Cranbrook Academy of Art.

Although the visual communications jurors—Richard Danne, George Finley, Gary Hanlon, Herbert Meyers, and Patrick Whitney—selected almost a quarter of the items they reviewed (36 of 131) for inclusion here, they felt the year produced very few examples of outstanding visual communications design. They noted, however, a general ripening, a leavening of communications design to the point where its overall quality is higher than ever before, leading the jurors to expect more of each entry.

Probably the most obvious example of this maturation is found in annual reports. Virtually all the annual reports seen by the jury were competently designed; they judged the five seen here the best of those. The jury was almost blasé in its acceptance of good annual report design. The state of that art has reached a point where, as one juror put it, "If a company doesn't have a good annual report, there is something wrong. You don't even need a designer to have a good-looking annual report. The standards are so widespread that a printer can turn out one that's competent."

In part the pervasive excellence of annual reports is a function of the money corporations are willing to spend on them. Annual reports will be available for a year, vying for attention in brokerage offices, in banks, in the homes of investors, on the desks of clients and suppliers. Corporations recognize the advantage of putting money into annual reports rather than into a TV spot seen for 30 seconds or even a magazine ad that runs for a month.

The money-design argument is, of course, an old one. Money does not ensure well-designed annual reports. "You can still do a lot of expensive, tasteless things," one juror noted. "And you can do things with one color that look fine, though the design may not be extravagant or elaborate." But the money spent on annual reports marks an attitude, an emphasis, a focus of attention that makes the difference between excellence and mediocrity.

Corporate reports raise other questions that apply to the entire spectrum of design. For example, once an existing design establishes a design format (in any design area: furniture, products, environments, etc.), is a subsequent design, following that format though equally well done, less exciting than the original? If as one juror suggested, "you took an annual report for IBM, done last year, and followed the same basic approach again this year, would you consider the new work exciting—even though it might be very well done?" Annual report design seems pinned to this totally appealing dilemma.

Consider another more fanciful dilemma. Designers spend their professional lives striving for design perfection. What if they achieved it? What if circumstances and talents raised the design of everything in America to the refined level of corporate reports? Would we find it an impossible world to live in? Would only designers be comfortable with it?

Slowly, corporate design awareness is seeping past the corporate report to engulf other corporate items. Design advantage can often be measured in cash, giving particular impetus to design for the marketplace. But noncorporate design is never far behind. Take the two corporate identity programs shown here: one for a chain of men's clothing stores, National Shirt Shops, the other for a nonprofit organization, Girl Scouts of America. Both are distinctive programs, both no doubt successful, but National Shirts can measure that success by reading its profit statements: sales are up an amazing 60 percent in those stores that have been redesigned to fit the new corporate image.

Five of the items seen in this section represent the J C Penney Co., though they do not necessarily represent Penney's corporate identity. The jurors felt, extrapolating from the 15 Penney's items they saw, that Penney's does not yet have a "complete design emphasis." Its packaging for instance, "though very well conceived, had nothing unique about it graphically." At the same time a juror noted: "Penney's is a company that has made a commitment to design and their program is really producing results now." Still the jury found no one "spectacular item" in the Penney's portfolio. Each Penney's item was judged separately, but the jurors wanted those selected to be shown together here.

Other corporate packaging seemed, in general, to appeal to the jury when it carried an artistic, sharp, arresting, usually colorful representation of the item contained. Most successful packaging today relies, of course, on this sort of graphic representation, on photographic glamorization. There's obviously a good deal of difference between seeing a noodle and seeing a photograph of cooked noodles garnishing a colorful plate of meat and vegetables on the Myojo package (page 118). Despite the effect of opulence the jury spoke out for simplicity in packaging. They liked J C Penney's packages for its 88¢ items precisely because the packages show the items (kitchen gadgets) and say 88¢ with no fuss or fanfare. "We are too often overpackaged," was a general feeling, as the jurors questioned the need for cardboard boxes to package the Dansk cutting boards (page 116) and the Copco knives (page 117).

But if the technical and artistic possibilities of photography are often fully exploited, those of typography are not. "There just wasn't a lot of creative typography," commented one juror. "The backbone of this whole visual communications field is type, and the biggest problem I saw was a confusion of product names and titles, along with lousy communication."

One typographic trend that suited the jury was the appearance this year of a style practiced by graduates of Yale and Philadelphia College of Art. It is the juxtaposition of widely and closely spaced lines of type. An example appears on the flash tags J C Penney's uses on its boys' shorts (page 130).

Choice of a compatible typeface can, of course, enhance a design. That Helvetica is the pervasive typeface of the seventies does not necessarily mean that typography relying on it is unimaginative, though designers often equate Helvetica's overuse with a lack of imagination. Helvetica, a sans serif face, produces the clean, uncluttered contemporary look (though Helvetica dates from the 19th century) that is today's fashion. But a typeface chosen to enhance a design's ambience can only help a design communicate.

The jurors were also distressed by the quality of the copy the typography has to support. Messages on packages or in brochures are too often sloppily written, contended at least one juror. "There seems to be a lot of good design but not enough good writing, good style," he said.

Nor did the use of symbols and corporate logotypes please the jury. Only one logotype is included here, Saul Bass's promotional logo for CBS's 50th anniversary. Symbols, the jurors felt, are being overdone and, more distressing than that, done lazily, without care. "A symbol doesn't necessarily solve a communication problem in one case just because it has worked successfully somewhere else."

Other areas of communication prompted more optimism. Posters, which are sometimes a pure graphic exercise, show the least drift toward stereotype. They offer more diversity of paper stock, color, typeface, and graphic approach than do other areas of communication. The Cranbrook New York trip poster (page 147), for example, is a collage of photography and type; the Bass on Titles poster (page 148) achieves its effect by a juxtaposition of film strips; and the Halleluiah poster (page 149) by a listing of block prints.

Packaging is the most represented discipline in this section with 12 examples; posters are next with 6 (2 of which are a series). Annual reports have 5 representatives and so does J C Penney's design program. There are 2 corporate identity programs, 2 catalogs, and 1 each of logotypes, pamphlets, calendars, and sales literature. Even out of context these items communicate.

Prototype Package Design of Uncontested Divorce, Incorporation, Wills & Trusts, and Bankruptcy Without a Lawyer. Designer: Layne, Salvo & Associates, Inc., Waltham, Massachusetts: Thomas R. Salvo, Michael R. Layne, designers; Harley Gordon, lawyer/writing and research.

Dansk thought its previous teakwood products packages provided too little excitement. The company wanted to stimulate impulse sales among its mainly affluent, urban, female (25- to 45-year-old) customers. The result is redesigned packaging with stunning photography showing the product in use and with front-panel typography stating succinctly what the product is and that it comes from Dansk. Back panels explain the properties of teak, show how to care for the product, and credit the designers. All the 13-prod-uct packaging now has a family look.

A couple of jurors raised an "over-packaged" cry, questioning the need to package products they felt could be displayed, sold, and carried home without graphics and cardboard wrappers. "It doesn't make sense to have a package around a solid block of wood," said one juror. The majority, however, felt the packaging helped explain the items' uses to consumers. Sizes range from 6 x 12 x 1" (15.2 x 30.5 x 2.5 cm) to 18 x 24 x 6" (45.7 x 61 x 15.2 cm).

Materials and Fabrication: high-saturation litho bond on coated boxboard (die cut); four-process color, 8 x 10" (20.3 x 25.4 cm) format photography; white typography; brown background.

Manufacturer: Dansk International Design, Mt. Kisco, New York.

Staff Design: J. Christopher Hacker, senior product manager, marketing; Jerrold Ross, vice-president marketing; George Ratkoi, photographer.

Consultant Design: Group Four, Inc.: David W. Eaton, principal; Douglas H. Terrell, staff designer.

Copco found that few sales personnel and few consumers know what makes a good kitchen knife. So Copco's new knife packages carry copy telling them. This information is organized in categories (the blade, the handle, the collar, the tang, the edge) and accompanies a drawing of the particular knives the package contains. Though some jurors felt knives need no packaging, that they are their own best sales device, the packaging, Copco maintains, helps distinguish its knives from those of competitors. Packages taper toward an envelopelike closure at one end. The cutting knife package measures 18 x 9½ x 1¼" (45.7 x 24.2 x 3.1 cm); carving knife package is 15 x 4½ x 1" (38.1 x 11.5 x 2.5 cm); steak knife package 10 x 8¾ x 1" (25.4 x 22.2 x 2.5 cm).

Materials and Fabrication: SBS laminated, E-flute corrugated paperboard (die cut and scored); print is SBS lacquer-laminate.

Manufacturer: Copco, Inc., New York.

Staff Design: Marc-Albert Passy, vice-president/design director.

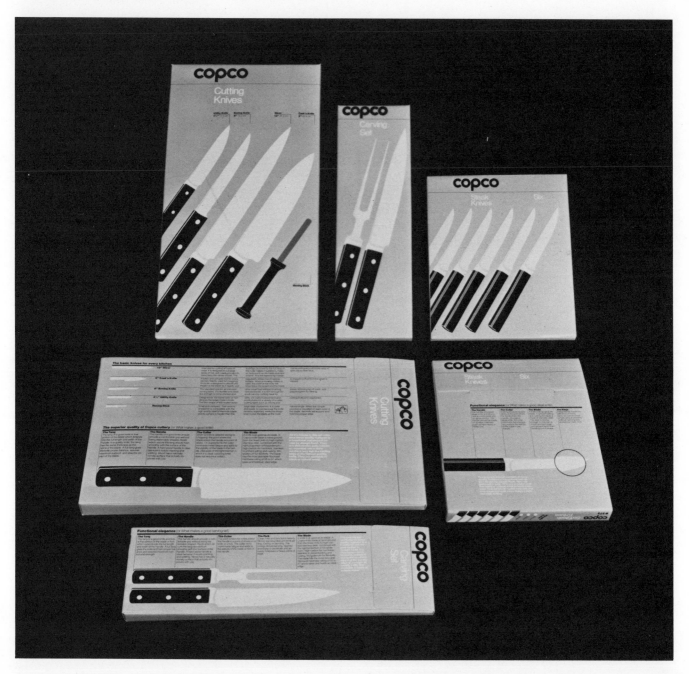

Myojo markets its ramen noodles in 3-ounce [5¼ x 4½ x 1¼" (13.3 x 11.5 x 3.1 cm)] and 9-ounce [8 x 3¼ x 3" (20.3 x 8.2 x 7.6 cm)] packages, whose outer wrappings are a combination of cellophane and polyethylene. On this surface is gravure printed a photo representation of the noodle's flavor—steak and noodles for beef flavor, for instance, a chicken leg and noodles for chicken flavor, and green vegetables and noodles for oriental vegetable flavor. Copy above and below the photo is angled at 23° so it yields more space to the photos and, according to the designers, is set in an italic face to suggest quickness of preparation. Packages for each particular flavor have a distinguishing color scheme (beef: maroon and red; chicken: orange and

yellow; oriental vegetables: blue and green). "It's absolutely beautiful," raved one juror. "The color is just beautiful." A paper band inserted in the soft 3-ounce package protects the noodles in overseas shipment.

Materials and Fabrication: polypropylene and polyethylene combination (heat sealed).

Client: Myojo Foods of America, Inc., New York.

Staff Design: Takashi Nishiyama, president; Katsuhiko Hayashi, marketing manager; Takashi Adachi, manufacturing and production manager.

Consultant Design: Robert P. Gersin Associates Inc.: Robert P. Gersin, program director, graphic design; Louis Nelson, project director; Carol Savage, project manager; Georgina Leaf, Casey Clark, graphic design.

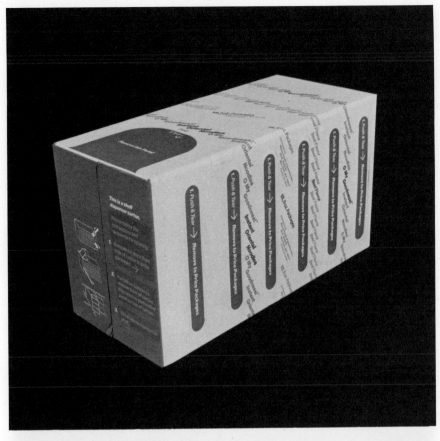

The soft packages enclosing Myojo's instant Oriental noodles are hard to stack on market shelves. They will not line up or stay put, preferring to slide and fall. So the designers came up with a shipping carton that becomes a dispensing-display unit on the market shelf. By removing a few die-cut strips from the carton the grocer can mark each noodle package without taking it out of the carton. He then removes a die cut front panel, exposing the noodle packages, puts the carton with the packages inside on the shelves, and lets customers help themselves. Diagonal carton graphics corresponding in typography and color to the graphics on the packages inside identify the carton's contents. Carton-opening instructions are in white type on an accent ground. Twenty-four 3-ounce packages are in a carton 9 x 9 x 10½" (22.9 x 22.9 x 26.7 cm). Eighteen 9-ounce packages are in a carton 9¾ x 18 x 7¾ (24.8 x 45.7 x 19.7 cm).

The jury worried a little about the grocer who will end up with an empty cardboard carton on his shelves and about the clerks who will have to read the carton's copy to know that its contents are to be left inside. "They have lots of boxes in the stores when they're stocking," recalled one juror who used to be a grocery clerk, "and they throw them on the floor, take out their knives, and cut. They don't bother to sit and read."

Materials and Fabrication: carton: bleached white E Flute corrugate, printed by flexography (die cut and perforated, glued).

Client: Myojo Foods of America, Inc., New York.

Staff Design: Takashi Nishiyama, president; Katsuhiko Hayashi, marketing manager; Takashi Adachi, manufacturing and production manager.

Consultant Design: Robert P. Gersin Associates, Inc.: Robert P. Gersin, program director; Louis Nelson, project director; Carol Savage, project manager; Robert P. Gersin, Pam Virgilio, graphic designers; Robert P. Gersin, Mo Khovaylo, carton designers.

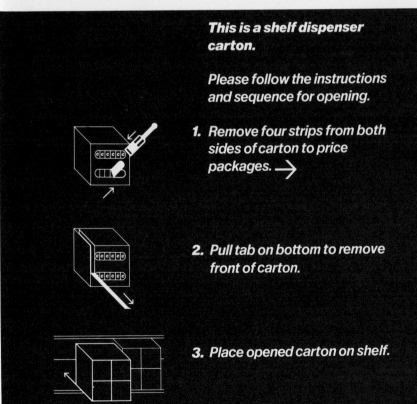

This is a shelf dispenser carton.

Please follow the instructions and sequence for opening.

1. Remove four strips from both sides of carton to price packages. →

2. Pull tab on bottom to remove front of carton.

3. Place opened carton on shelf.

The Panshake Pancake Mix package leads a quadruple life. It is both a container for the mix and a mixing bowl for the consumer. Lined with foil, the carton holds the water a consumer adds to the mix within the carton. It then becomes a shaker and a pitcher from whose taper top the mixed batter is poured into a frying pan or onto a griddle.

The designers feel the carton's white background helps it stand out on the market shelf. Bold red letters slanted across the carton's peak and sides announce the name: Panshakes Pancake Mix. Blue letters just beneath this logo spell out how the product may be used: "Just add water to the carton." And a photo showing a hand pouring batter from the carton into a pan on a side panel reinforces the lettering.

Materials and Fabrication: Tapertop paperboard carton .022 CCNB with Coex film liner. Printed in five colors plus varnish in offset lithography.

Client: The Pillsbury Company, Minneapolis, Minnesota.

Staff Design: Joel E. Burke, group marketing manager.

Consultant Design: Gerstman + Meyers Inc.: Richard Gerstman, design supervisor; Larry Riddell, Ed Silenski, designers.

On the market in the Southeast since 1965, KICK, Royal Crown Cola's caffeinated, citrus-flavored soft drink, came in a red can with a picture of a kicking mule on it. To take KICK nationwide Royal Crown ordered a packaging redesign, and the designers came up with a simple logo and color scheme easily applicable to all KICK packages—cans, bottles, labels, and cartons. The logo, placed on a white background, is "KICK" in red letters (with a large green dot over the "i") underlined by a diagonal green gradient meant, according to the designers, "to imply the vibration inherent in KICK."

Materials and Fabrication: Cans: white-coated steel (printed by lithography); bottles: glass (labels printed offset); plastishield bottle (printed flexographically); eight-bottle carrier carton: white chipboard (printed flexographically).

Client: Royal Crown Cola Company, Schaumburg, Illinois.

Staff Design: Laurence Atseff, marketing director.

Consultant Design: Gerstman + Meyers Inc.: Herbert M. Meyers, design supervisor; Juan Concepcion, design director; Rafael Feliciano, designer.

J.A. Wright & Company recently gave the graphics on all its metal-polish containers a similar look. An arch of dots, which now radiates above the name of the cleaner (Wright's silver cleaner and polish, Wright's brass cleaner and polish, etc.) is meant to create a shining image. Not so effective aesthetically when used on paper labels or paperboard boxes, the graphics are outstanding when printed on metallic containers whose natural gleam shines through the radiating dots. Most effective, the jurors felt, is the 7-pound (3.1-kilogram) institutional size silver cleaner can. Here the radiating dots, printed directly on the can's gleaming tinplate, have scope not possible on the smaller container and a natural brilliance not allowed by paperboard. "It says glowing shine," commented one juror.

Materials and Fabrication: 7-lb institutional silver polish can: tinplate with blue lithography.

Manufacturer: J.A. Wright & Company, Keene, New Hampshire.

Consultant Design: Selame Design: Joseph Selame, president, chief designer; Robert Selame, senior designer.

Rhodes markets its steel wool pads under two brand names: "Beaver" and "Sunray." Before redesign, all grades of steel wool had been presented in somewhat similar packages. A customer could discover what grade his package held only by referring to a sometimes confusing numbering system. To make the various grades easily recognized and selected, the designers replaced the numbering with graphics. Now, for instance, a steel wool grade for paint and finish removal is packaged in a cardboard box with a stylized drawing of a table corner on it. An automotive grade carries a drawing of a car's front fender. In each package a round, see-through, cutout lets a customer catch a glimpse of the steel wool inside. Print is contemporary, brightly colored, and easily read.

Client: James H. Rhodes Co., Des Plaines, Illinois.

Consultant Design: Design West, Inc.: Ted Piegdon, creative director; Rich Hurtado, Tom Suiter, designers.

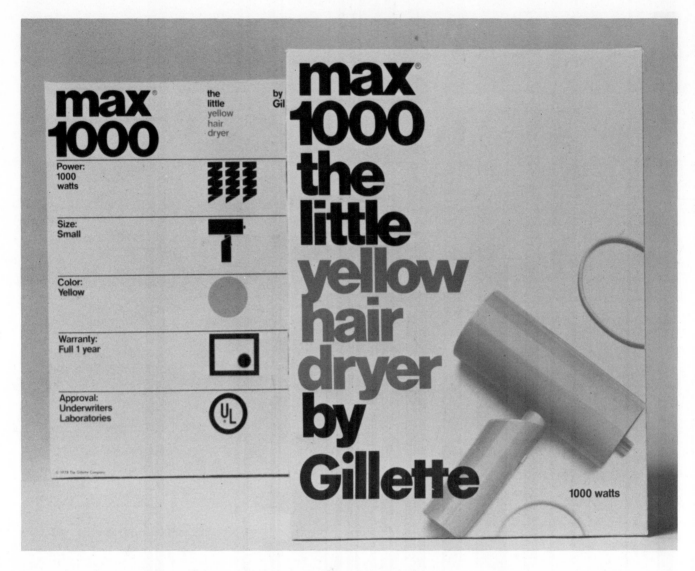

The "little yellow hair dryer" is Gillette's subtitle for its Max 1000. Selling for $12.99, it competes with off-brand dryers, so the package emphasizes the company name, Gillette, while staying simple and straightforward. It is perhaps no coincidence that the box front ends up looking like a book jacket: book jackets have long presented a maximum amount of information with minimum fuss. But it is not *too* straightforward. "It's a little playful, which I like," commented one juror. On the package's back is a list of the dryer's major advantages (one-year warranty, UAL approval, etc.), each given equal space and each made noticeable by an accompanying graphic symbol. "Organized" is the way one juror described it.

Materials and Fabrication: bleached sulphate, laminate paperboard (die cut); colors; black and two yellows.

Client: The Gillette Company, Boston, Massachusetts.

Consultant Design: Morison S. Cousins + Associates Inc.: Johann Schumacher, graphic design.

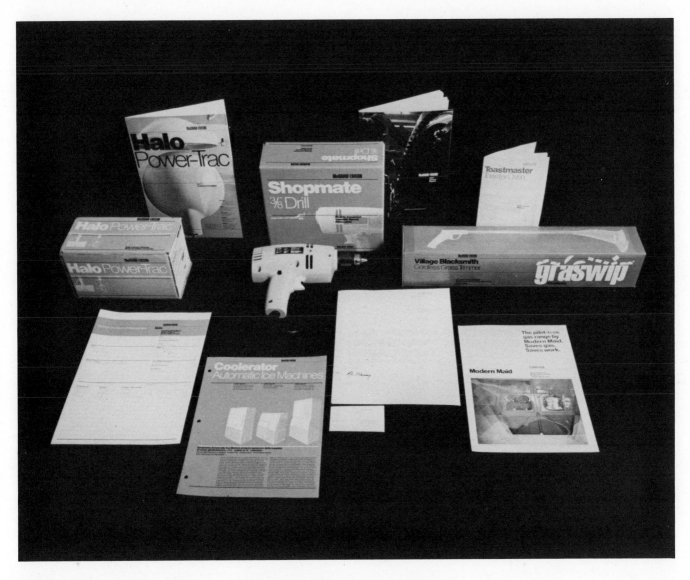

McGraw-Edison makes a host of consumer and industrial items, such as cordless grass trimmers, automatic ice machines, and power hand drills, which it sells to retail outlets (consumer products) under several brand names or directly to distributors and users (industrial products). McGraw-Edison's redesigned packaging system links all these items, identifying each closely with McGraw-Edison and accordingly with one another.

Though the packages vary widely in size and color, identity is stressed by typography: the uses of similar typefaces, similarly positioned on the packages (arranged in softly lined sections). The jury was unanimous in its praise of the system, calling it "outstanding" and "effective," if not unique.

Materials and Fabrication: boxes: natural Kraft corrugated, bleached Kraft corrugated, sulfite board folding cartons; printing: offset and flexographic.

Client: McGraw-Edison Company, Elgin, Illinois.

Staff Design: Jeff Rogalski, manager of design.

Consultant Design: Anspach Grossman Portugal Inc.: Eugene J. Grossman, design director/designer; Sam Shoulberg, artist.

GAF markets more than 100 items under its GAFMED label—mostly x-ray film and chemicals—sold directly to industrial and institutional clients. To make it obvious that all these products come from the same manufacturer, designers Danne & Blackburn evolved a graphic system using color (red, white, and black), type (Helvetica), and the GAFMED logotype, which has a medical cross as the counter in the letter D. The strong, bold red GAF-MED logo appears in a rectangular white section at the top of the front of the package, label, or literature—just above a red rectangle which carries the name of the particular product in black type. GAF found that its GAF-MED products now appear related not only to each other but to the parent company as well. And the family appearance saves the company money, for it no longer has to print a disparity of labels and package covers.

Client: GAF Corporation, New York.

Staff Design: Paul Miller, design director.

Consultant Design: Danne & Blackburn, Inc.: Bruce Blackburn, designer.

Even J C Penney's automotive batteries now have a Penney's look. All information needed by a customer in selecting a battery is displayed right on the battery itself (size, cold crank rating, warranty, a warning about battery acid). The number of years of the warranty is printed in red. All other printing is white against the black battery cases. Printing is done on 6-ml vinyl labels, resistant to heat, oils, and battery acid, which stick to the battery case with a rubber-based adhesive.

These labels, once stuck to the batteries, are laminated with 1 ml of clear polypropylene. Penney's battery line has 16 batteries of various sizes and crank ratings for use in almost any car. But some of the labeled information can be used on all batteries, thereby cutting costs. The designers also devised a date-car-model code which goes on each battery to keep customers from falsifying warranty information.

Materials and Fabrication: labels: vinyl used with a rubber-based adhesive; printing: flexography on a webtron press.

Client: J C Penney Company, Inc., New York.

Staff Design: Ron Pierce, senior designer; Ira Gaber, project manager.

Each year J C Penney displays—stacked, hanging, or mixed randomly in bins—a new group of about 75 gadgets, which it sells for 88¢ each. These items are selected by a Penney's buyer in Taiwan, and all mechanicals for their packaging and the actual package printing are done there. Offered as impulse purchases in Penney's stores, the items are meant as a temporary promotion.

Now, as part of Penney's corporationwide design emphasis, their packaging is standardized. Before, the items often came in manufacturer-supplied packages, covered with renderings of French chefs and dancing vegetables and offering no coherence in color or typeface. Packages now, though of several kinds and sizes, are all offset printed in yellow and orange on white and conform to federal standards, displaying succinct information about the contents. Actually the packaging is minimal, in each case allowing the product to show and thereby letting the consumer focus on what he or she might be buying. This simplicity impressed the *Design Review* jury. "Simplicity is what makes it stand out," said one juror of the packaging. "It communicates very well," said another. "The economy of the whole presentation is admirable."

Offered as packages are blister cards, cardboard headers, circular inserts, and a couple of miscellaneous constructions. The Penney's designers also developed guidelines and partial mechanical art so foreign manufacturers can package according to Penney's standards.

Materials and Fabrication: paperboard inserts, headers, and other specialized construction in combination with blister packs, shrink wraps, and plastic bags; two-color offset printed.

Client: J C Penney Company Inc., New York.

Staff Design: Marjorie Katz, creative director; Carol Zimmerman, designer.

J C Penney's sells two lines of leotards under its own label: a basic line, used for dance, exercise, and casual wear, and a fashion line, which can be used, in addition, for evening wear and swimming. Previous basic leotard packaging consisted of a paperboard envelope with one small window. The new package, which, Penney's claims, cuts packaging costs 27 percent, is simply a paperboard printed with a black-and-white product photograph and wrapped, with the leotard, in a polypropylene bag, making it easy for a customer to see the merchandise.

This bag closes with a double-back tape, so a customer can open the package, try on the leotard, and then repackage it. For its fashion line packages Penney's uses full-color photography and on packages for both lines four-color printing. Each printed board is scored at the bottom to fold back, forming a rectangular base on which the package can stand when on display. An OCR label applied to the package's upper right-hand corner carries information about color, fiber, and size. Package size: 9 13/16 x 6½ x 3" (25 x 16.5 x 7.6 cm).

Materials and Fabrication: 12-point scored J-board, packed in a polypropylene bag. Printing: four-color, eight-up offset.

Client: J C Penney Company, Inc., Packaging Department, New York.

Staff Design: Marsha C. Adou, package designer; David Law, design director; Marjorie Katz, creative director; Roxane Pandya, packaging copy writer; Marilee Witt, package engineer.

Photographer: Avedon Studios: Gideon Lewin.

129

Part of J C Penney's aggressive, pervasive redesign of everything that touches its corporate image, these flasher tags for boys' shorts are indicative of the care that is going into the program. The tags are 2½" (6.4 cm) wide by 3¾ (9.5 cm) high with a 1¼" (3.1 cm) flap that hooks over the shorts' waist bands. Each tag adheres to engineering specifications, using a 1¼" module, which allows tags to be easily developed for new shorts, which will be added to the existing line.

Materials and Fabrication: tags: 10 point SBS paper, two-color offset printed.

Client: J C Penney Company, Inc., New York.

Staff Design: Marjorie Katz, art director; Steven Schnipper, designer; David B. Law, package development manager.

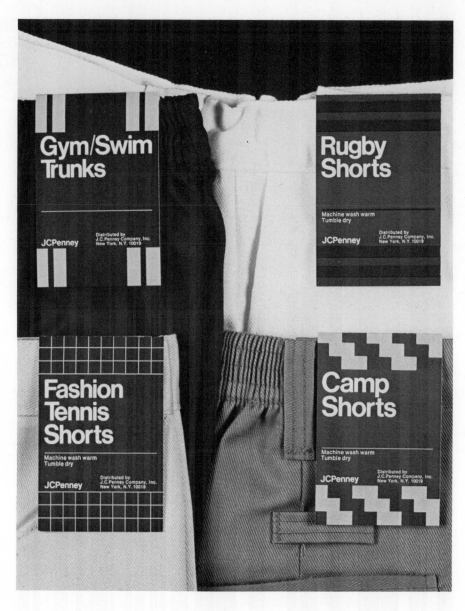

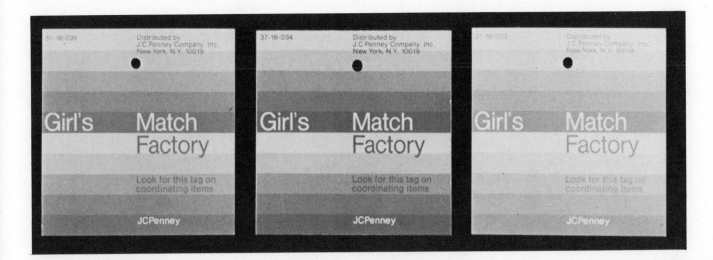

With the aid of these redesigned, color-coded 2¼ x 2¼" (5.7 x 5.7 cm) tags, children and their parents can locate coordinated items, socks and shirts that go with particular shorts, for instance, in Penney's stores. The new tags replace ones that had a more cluttered design showing a drawing of a factory. Each tag is printed in gradations of a single color. Unanimous in their praise of these hang tags, the jurors liked their particular use of color and the way the tags fit into Penney's overall design.

Materials and Fabrication: tags: paper printed with one-color lithography.

Client: J C Penney Company, Inc., New York.

Staff Design: Bill Bonnell, manager of packaging development.

Shape Op® is a prototype seamless package, formed from a poly-foil-poly lamination blank; its cover is a polyethylene overcap. Container Corp. claims that its continuously connected base and sidewall and its tight cover, achieved by heat sealing a membrane to the carton's polyethylene upper rim, will give superior protection against sift, moisture vapor, infestation, or tampering. Uses should include packaging powdered products (such as cocoa, flour, and powdered milk), granulated products (such as drink mixes and cleansers), or solid products (such as hardware, toys, or candy). Stored or displayed on shelves, where they take up 25 percent less space than round containers, their graphics are easier to read. The package's overcap offers easy access and is preferred by consumers, who, studies show, are not satisfied by shaker caps, metal plugs, or holes covered by adhesive patches. Print can go on the package's seamless bottom; and package height, Container Corp. maintains, can be changed with only minor machine alterations. Shape Op® container sizes range generally from 2 x 3 x 2.5″ (5.1 x 7.6 x 6.4 cm) to 5 x 7 x 8.5″ (12.7 x 17.8 x 21.6 cm).

Materials and Fabrication: base package: board: .024″ CT80NB soap back liner lamination (½ mil poly, ½ mil foil, 1 mil poly; finished board caliper: .026″; closure: overcap: low-density polyethylene; rim: high-density polypropylene; membrane: .0015 foil with 3 # heat seal coating.

Manufacturer: Container Corporation of America, Carton Division, Oaks, Pennsylvania.

Staff Design: Michael A. Kipp and Richard L. Bell, senior industrial designers; Robert M. Redding, project engineer.

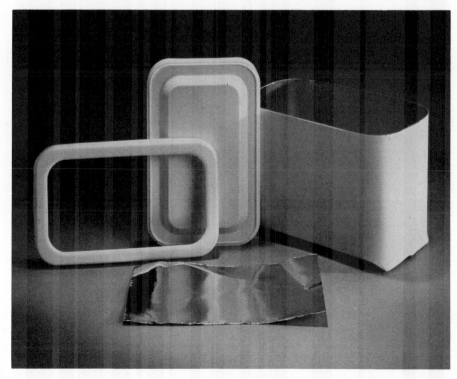

The *Design Review* jurors cited Container Corporation's plastic tank sales literature for its high quality of design, its inviting clarity, and its aesthetic appeal. "Refreshing for corporate work," they called it. The literature is meant both as a mailing piece and as an aid to sales personnel. The polyethylene tanks described in the literature are used for storing, mixing, handling, and compounding chemicals, industrial cleaners, soaps and detergents, pigments and dyes, food products, and waste disposal, among other things. The tanks are products not usually touted so nobly. Pages measure 8½ x 11" (21.6 x 27.9 cm) and are printed in two colors.

Materials and Fabrication: paper, printed by offset lithography.

Manufacturer: Container Corporation of America, Plastics Division, Chicago, Illinois.

Staff Design: Bill Bonnell, designer; Bob Best, writer.

Container Corporation's 1978 promotional calendar opens like a pinwheel, displaying a rainbow of 14 colors. In its pinwheel form it sits on a desk decoratively but, as the *Design Review* jurors pointed out, "impractically." What they meant was that in its pinwheel form, the calendar is practically unreadable. If that bothers you, you can set it up in its clear plastic holder, displaying one month at a time. Folded the calendar measures 4 x 4 x 1.5" (10.2 x 10.2 x 3.8 cm).

Materials and Fabrication: calendar: paper board laminated to corrugated paper; clear holder-box (injection molded).

Manufacturer: Container Corporation of America, Chicago, Illinois.

Staff Design: Bill Bonnell, manager of design; Bill Kaulfuss, structural design consultant.

Arranged in a packet of 11 single sheets and brochures—some having as many as six pages, formed by folding larger sheets to the 8½ x 11" (21.6 x 27.9 cm) size of the single sheets—Raychem's Corporate Identity Guide comes in a paper-jacket folder of Raychem red. The guide grew out of the company's expansion here and overseas, a growth that strained the ability of an informal corporate identity program to cover all facets of corporate design in all Raychem offices and plants. Though the corporate identity crisis produced by expansion was low

key, the need for codification of the corporate design program did become apparent, and the resulting packet of directions can lead any Raychem office manager or purchasing agent through the complexities of keeping the Raychem name straight and prominent. Most such guides are designed for design professionals to use as references in applying the corporate name to the welter of corporate signs, cartons, vehicles, purchasing forms, order blanks, and so on. Not so Raychem's. Instead it is meant to be understood and followed by anyone in

ordering stationery or furniture, in typing memos or putting logotype decals on corporate trucks. The jurors not only liked the design standards set forth in the guide but praised also its clarity and ease of use.

Materials and Fabrication: paper: 100 # Vintage vellum; printing: two- and four- color offset.

Client: Raychem Corporation, Menlo Park, California.

Staff Design: John Rieben, design director; Richard Klein, Richard Stanley, Nancy Long, Ann Kortlander: designers.

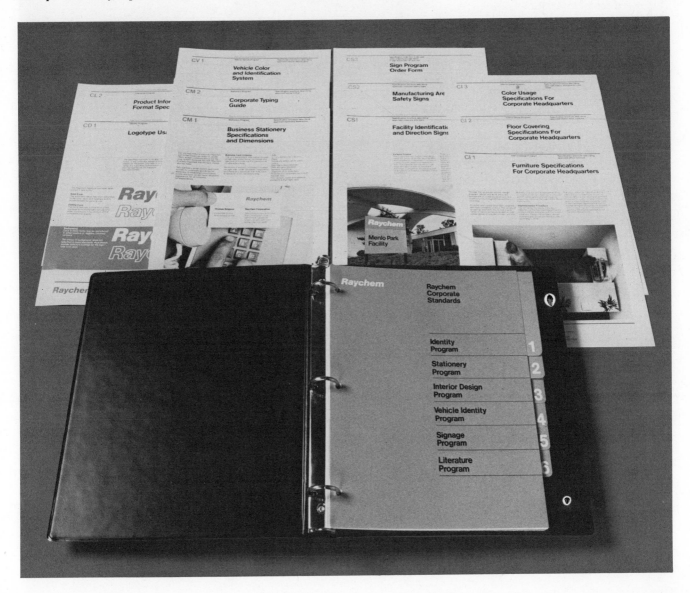

In a series of five booklets IBM explains what it's like to work for IBM and what employees do while pursuing specific IBM careers (in marketing and systems engineering, in science and programming, in administration and finance). Each 20-page booklet has the same format and a cover composed of 12 square color photos of IBM people at work. (Front and back covers are identical, except for "IBM" in striated white letters at the upper right of the front cover.) Inside are larger photos, some of them repeating the smaller photos on the cover, and captions and text explaining the IBM message ("The first bulwark against the impersonality so often found in business organizations lies in one of IBM's three fundamental beliefs: *respect for the individual . . .*"). These booklets are used in IBM's recruiting program. A potential employee is given a copy of the general corporate booklet and a copy of the booklet pertaining to his particular career. All are 7½ x 11" (19.1 x 27.9 cm) with saddle-stitched bindings.

Materials and Fabrication: paper: Northwest Quintessence gloss, 100 # ; printing: four-color offset lithography.

Client: IBM Corporation, Armonk, New York.

Staff Design: Jerry Blood, project coordinator.

Consultant Design: Danne & Blackburn, Inc.: Bruce Blackburn, art director and designer.

Robert P. Gersin's 101-page catalog for the 1977 American Institute of Graphic Arts (AIGA) packaging exhibit shows items selected for inclusion by a jury of four designers, a marketing specialist, a retailer, and a consumer. On the catalog cover is the show's logo (also designed by Gersin), a variation in blue of the universal price code: that block of varying width black lines that appears on most packages nowadays. Gersin's variation has the lines bent in an obtuse angle like the cover of an open matchbook. Set into the lower edge of these lines is the date of the exhibit opening. Offsetting this logo, along the left edge of the catalog's front cover (with echoes along its spine) are the orange letters "a i g a," 1⁄8" (3.2 mm) high, spaced 1½" (3.8 cm) apart. Gray letters spell out the exhibit's (and the catalog's) title: "What's Real in Packaging?" Inside are black-and-white photos of the show's packages, most one to a page, surrounded by an admirable amount of white space, and accompanied by a block of text in which each package designer explains his design. The catalog designers also worked out a code of blue lines spaced along page margins to show which AIGA jurors voted for each entry. Catalog size is 7 x 10 x ¼" (17.8 x 25.4 x 0.6 cm).

Materials and Fabrication: cover: Champion 80# Wedgewood paper, dull; interior pages: West Virginia Paper Co. 80# Sterling Matte; type: PMS 408, PMS 151; binding: Flexico adhesive process.

Client: American Institute of Graphic Arts, New York.

Staff Design: Nathan Gluck, assistant director.

Consultant Design: Robert P. Gersin Associates Inc.: Robert P. Gersin, program director; Robert P. Gersin, Ronald Wong, Kenneth R. Cooke, graphic design.

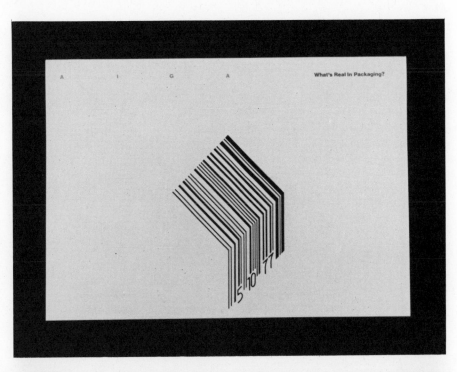

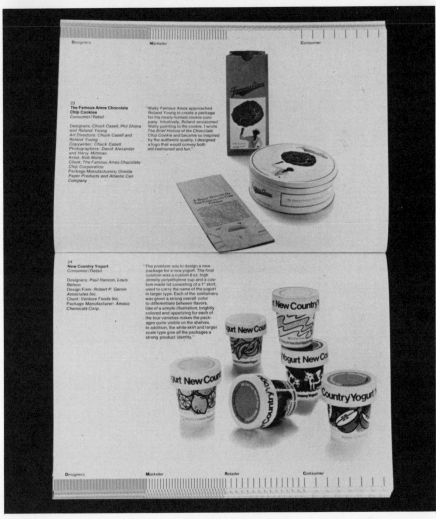

It's not that the Girl Scouts of the United States of America have a new image so much as that they have a consolidation and clarification of the old one. Until October 25, 1978, when the Girl Scouts of the U.S.A. displayed their Saul Bass/Herb Yager-designed logotype to the Girl Scout convention in Denver, the Girl Scouts had been identified by a welter of symbols, mostly based on the trefoil or on groupings of the Girl Scout name. The Bass/Yager logo combines a symbolic portrayal of three young women's head in profile (two white, one black), done in green and white and underslugged by "Girl Scouts" in blue. Though these colors may not be reversed, the entire logo can be printed in a single color, "provided," says the Girl Scouts' graphic guidelines manual, "the color selected offers sufficient contrast to the background on which it is placed." By using the image of young women in the logo, the organization stresses its difference from the Boy Scouts and, by the force of the aesthetics, gives itself a more contemporary, attractive, less tightly institutional image. The new logo will appear widely: on cookie packages, T-shirts, posters, decals, badges, and all print material used by the Girl Scouts of the U.S.A.

The jurors initial reaction to the logo was that the smooth flowing masses of hair made the girls look more adult than girls elegible for the program. But the jury decided that "although the image isn't little-girl-like, it probably reflects something the girls want to become, something they aspire to." And one juror, though finding the design only fair, thought that getting a new logo approved by the institutional bureaucracy of the scouting organization was a triumph.

Client: Girl Scouts of the United States of America, New York.

Consultant Design: Saul Bass/Herb Yager and Associates: Saul Bass, art director; Art Goodman, Vahe Fattal, designers.

The new face of Girl Scouting

GIRL SCOUTS

GIRL SCOUTS

NEWS

GIRL SCOUTS

Girl Scouts of the United States of America
830 Third Avenue
New York, New York 10022

GIRL SCOUTS

Girl Scouts of the United States of America
830 Third Avenue
New York, New York 10022

GIRL SCOUTS

Girl Scouts of the

Ely List
Associate Director
Public Relations

GIRL SCOUTS

Girl Scouts of the U.S.A.
830 Third Avenue New York, NY 10022
212 940-7802

National Shirt Shops sell moderately priced men's clothing in 280 retail stores located throughout the U.S., mostly in shopping malls. Their corporate identity program extends to all the shops' elements and has as a central focus a unifying symbol; a fibrous letter N, comprised of five Ns, each overprinted slightly to the right of its predecessor. Suggesting a woven cloth pattern, this symbol greets shoppers as a transilluminated sign in front of every National Shirt Shop and is repeated inside as a 7' 6" (2.3 m) square entrance wall graphic. Throughout the program this N's color varies. (The jurors pointed out that its color works whatever it may be.) Sometimes all elements of the fibrous N are the same color; sometimes each of the five fibers has a separate color. It is this logotype's strength that it is equally recognizable as the 7' 6" wall graphic and on a 1¼ x 1¼" (3.1 x 3.1 cm) clothing label. In between the program includes design of sales forms, shopping bags, price tags, hangers, point of sales devices, sales personnel buttons, gift certificates, stationery, garment overwraps, ordinary paper bags, and even a gift certificate envelope seal. The jurors liked the program's flexibility within "a rather consistent image" but did not praise all its elements.

Materials and Fabrication: poly bags: polyethylene film (heat sealed); paper bags (glued), offset printed; overwrap: buff-colored polyethylene film (heat-sealed and perforated); sales slips: NCR paper, offset printed; labels, woven; stationery: mimeo printed, PMS-348 type.

Client: National Shirt Shops, New York.

Staff Design: Edward Marks, president; Steven Marks, assistant to the president; Daniel Markowitz, vice-president and general merchandising; Jules Kovner, advertising manager.

Consultant Design: Robert P. Gersin Associates Inc.: Robert P. Gersin, Louis Nelson, program directors; Lee Stout, project director; Kenneth Cooke, Alea Garrecht, Casey Clark, graphic design.

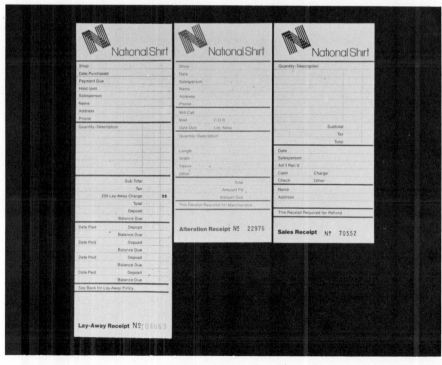

A Celebration of 50 Years

Saul Bass calls the logotype he designed for CBS's 50th anniversary a trademark. Just when a trademark became a logotype, or more commonly a logo, and when a logo came to refer not just to a slug of type but to an entire corporate symbol are matters for historians to ponder. But it can probably be argued with some assurance that for most of CBS's 50 years whatever symbol it identified itself by was called a "trademark." In a year in which RCA records went back to its old trademark of a dog cocking an ear to hear his master's voice (no one would call *that* a logotype), logotypes may be dissolving back into trademarks.

Bass's CBS trademark, meant to promote CBS's 50 years on the air, appeared on a range of items—stationery, news releases, press kits, and TV film—and it was used as the network identifying mark during TV station-breaks. One juror remembered it on the air: "It was a neon tube that animated itself. It was very lively and showbizzy." With its strong, semi-continuous flowing lines, it conveys a sense of motion. Whether the trademark's finger is pointing into the past or the future remains vague despite the subhead "A Celebration of 50 Years." One might argue that CBS is as much looking to the next 50 as recalling those just past.

Client: CBS, New York.

Consultant Design: Saul Bass/Herb Yager and Associates: Saul Bass, designer and art director.

Simple and colorful, the designs for Raychem's corporate posters (a series of four) show a fragmented, multi-colored circle, looking the way the sun might in a Navaho weaving, on a background that is usually some solid shade of brown. Meant as wall decoration in offices and buildings throughout the corporation, the posters' typography says only "Raychem" and "Responding to the needs of the Energy industry" (or Process industry, Electronics industry, or Tele-communications industry). According to the designers the design is meant to communicate the idea of energy and electronics. They designed four posters, one signifying each of the company's four new divisions (Energy, Telecommunications, Processes, and Electronics), which were consolidated last year from eight the year before. Raychem tries to put out a new set of posters each year. All four measure 41 x 29⅝" (104.1 x 75.2 cm).

Materials and Fabrication: paperstock: Shasta Coverway; printing: seven-color, silk screen.

Client: Raychem Corporation, Menlo Park, California.

Staff Design: John R. Rieben, director of design.

Raychem

Responding
to the needs of the
Telecommunications
Industry

Raychem

Responding
to the needs of the
Process Industry

143

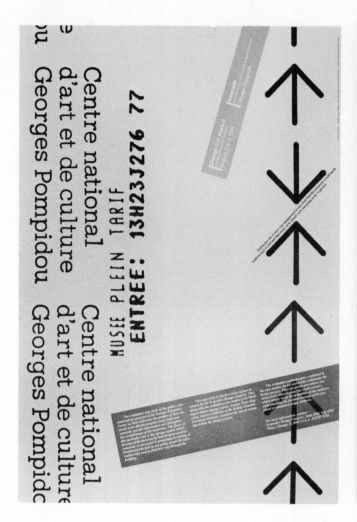

The International Council of Graphic Design Associations (ICOGRADA) held its first North American Congress at Northwestern University in 1978. To publicize case studies worked up for evaluation in the Congress's sessions, ICOGRADA produced four posters, three of which are included here. Posted at the Congress site and sold as mementos, the posters touted these case studies: the Washington Metro sign system (a highly praised design which, however, fails to communicate all the information to all the people); Paris's Georges Pompidou Center's public communication system (a design which successfully communicates the institution's image but not the information needed by visitors); and the design of *Sesame Street* magazine (an evaluation of how children see). Each poster is 23⅝ x 33'' (60 x 83.8 cm), printed in four colors on white stock: the Pompidou poster has a bright yellow background; the Washington Metro poster's background is black; and the *Sesame Street* magazine poster has a sand-colored background. The drawing of the cat is by Ania. That's the way she sees cats.

Materials and Fabrication: paper: Nekoosa opaque offset vellum; finish: white basis 80#; printing: four-color offset lithography.

Client: ICOGRADA Chicago Congress.

Staff Design: Robert Vogele, project director, Patrick Whitney, design director and designer of the Washington Metro case study poster.

Consultant Design: Francois Robert, designer, Pompidou poster; Ania and John Greiner, designers, Sesame Street poster.

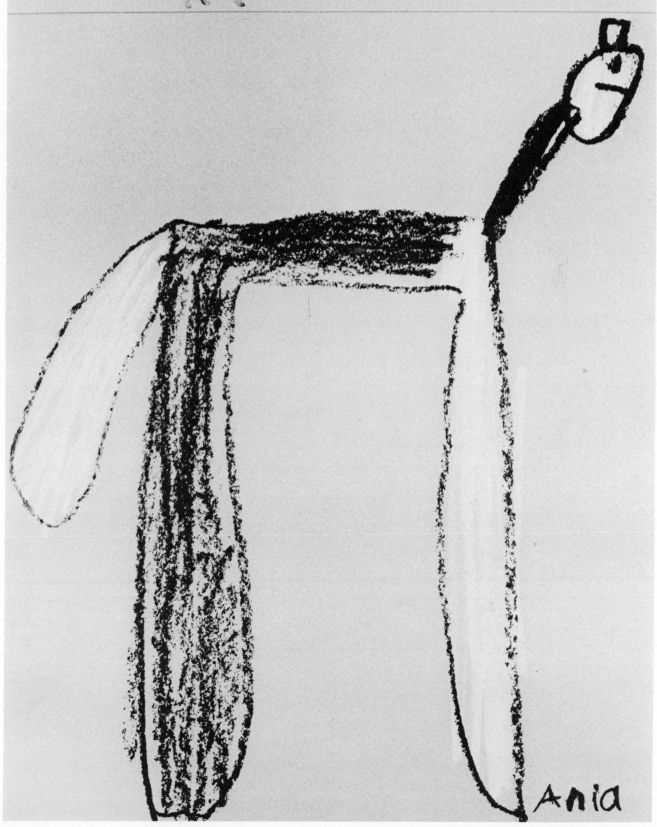

Design that **works!**
Evaluating design for the future
August 3 to 7, 1978

icograda
international council of graphic design associations

Chicago Congress

The evaluation case study of Sesame Street Magazine explains how a monthly magazine with critical time restrictions can effectively use formative evaluation. Through a simple testing process and a growing body of knowledge about design conventions that can be understood by young children, the evaluation team helps some of New York's best designers/illustrators match their images to the way children see.

This case study is one in a series of four developed for the Icograda Chicago Congress. This, along with the evaluation case studies of the Washington Metro signage system, the public communications program of the Georges Pompidou Center, and an exhibition of the British Columbia Provincial Museum, examines the role of evaluation within the design process.

The evaluation methodologies, explained in the case studies, do not necessarily systematize the design process. They are proven procedures borrowed from education and the social sciences which simply verify the probability of the audience understanding the intended message of the proposed or completed design solutions.

For more information, contact:

Icograda Chicago Congress
Suite 2900, One IBM Plaza
Chicago, Illinois 60611 USA
Telephone 312/787-2228

Cranbrook offers a two-year master of fine arts program in two- and three-dimensional design (industrial and graphic design, exhibit and furniture design, and interior architecture), and this poster is meant to spread that news, to both potential students and potential employers of the school's graduates. The school sends the poster to undergraduate design departments, design educators and journalists, museums and selected professional designers.

Its textual message presented in blue type is succinct, its overall graphic impression abstract. It shows a slate roof section (of one of the department's two Saarinen-designed buildings) floating inside another slate roof section. Essentially the 22 x 28″ (55.9 x 71.1 cm) poster is black and white, but its tones are given warmth and its abstraction structure by three strong solid-colored lines—red, yellow, or green—one at the poster's bottom edge and two outlining the top and bottom of the slate roof within the slate roof.

Materials and Fabrication: 100 # Mead dull enamel; printing: five-color offset lithography.

Client: Cranbrook Academy of Art, Design Department, Bloomfield Hills, Michigan.

Consultant Design: McCoy & McCoy: Katherine and Michael McCoy, graphic design. Photographer: Richard Sferva.

Meant to publicize Cranbrook's graduate design department's New York City trip of March 1978, this poster is also a departmental promotion piece. Mailed to New York designers and schools, it has been helpful, Cranbrook's design department claims, in attracting applicants. The poster, 17 x 34″ (43.2 x 86.4 cm), accents its own and New York's verticality by showing vertical sections of three well-known New York buildings (Empire State, Chrysler, and Citicorp Center). The yellow of a couple of taxis careening among the buildings at the poster's bottom is its only color. All this is set in a newspaper-type format, with "New York City" slugged at the top in *The New York Times's* font and the trip schedule set in long newspaper-like columns amidst the buildings.

Materials and Fabrication: paper: dull enamel stock; printing; two-color offset lithography.

Client: Cranbrook Academy of Art, Design Department, Bloomfield Hills, Michigan.

Staff Design: David Sterling, editor and designer: Katherine and Michael McCoy, art directors.

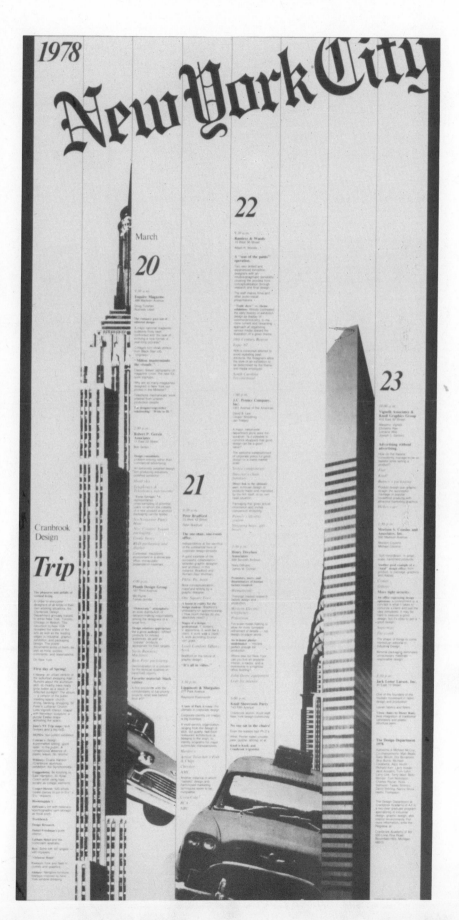

This promotion piece is for a 35-minute film in which Saul Bass discusses, specifically, the relations between 10 of his movie titles and the stories they illustrate and talks, more generally, of the title as film form. The "Bass on Titles" poster shows juxtaposed film strips from eight of these Bass titles. Best known among his titles are those for *Man with the Golden Arm, Seconds, In Harm's Way, West Side Story, It's a Mad, Mad, Mad, Mad World, Big Country, Grand Prix*, and *Walk on the Wild Side*.

Client: Pyramid Films, Santa Monica, California.

Consultant Design: Saul Bass/Herb Yager and Associates: Saul Bass, art director and designer.

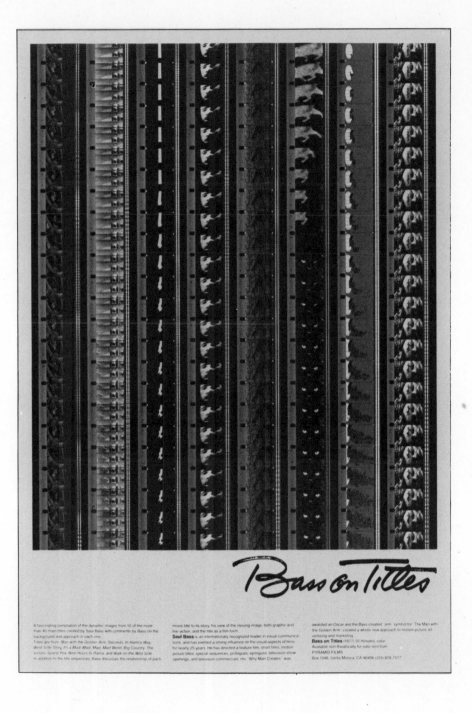

It was the Christmas holidays and Gary Hanlon & Associates had just opened new offices in Boulder, Colorado, so Hanlon decided to express his joy in both events by designing a poster. In five series of 4″ (10.2 cm) square boxes Hanlon represents the beginning of things—a star, a snowflake, a Christmas tree, a reindeer, and an angel. Each series has its own color—blue, orange, green, red, and purple—all on a white background; and the image in each series of four boxes starts with a line or two and develops across the poster. At the bottom in ⅝″ (1.6 cm) blue letters spaced 1½″ (3.8 cm) apart is one word: Halleluiah.

Materials and Fabrication: 60# gloss finished paper; printing: six-color offset.

Designer: Gary Hanlon, Boulder, Colorado.

Cluett is essentially a manufacturer of men's and women's clothes (Arrow is a Cluett label), but it also makes machinery for the apparel industry, operates three chains of Midwestern retail stores and a warp knitting facility, and has worldwide income from brand name, processes, and technology licensing.

The company illustrated its 21-page 1977 annual report with still lifes of the company's clothing, arranged as if it were just being taken out of the box. Colors are muted, soft blues, golds and browns, and whites. The effect is clean and the financial statistics are arranged on white stock with bold headings for easy digestion.

Materials and Fabrication: 80 # Cameo gloss; printing: sheet-fed offset.

Client: Cluett, Peabody & Co., Inc., New York.

Consultant Design: Cook and Shanosky Associates, Inc.: Roger Cook and Don Shanosky, designers. McGhie Associates, Inc.: Bruce McGhie, financial consultant.

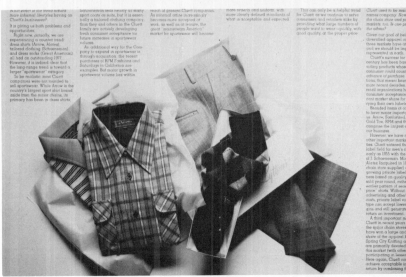

The Philadelphia-based investment banking firm, First Boston Corporation, operates worldwide. In 1977 the corporation opened its ninth international office, in Singapore. This office, in Singapore's DBS Building, appears on the cover of First Boston's 1977 annual report, photographed across the ornate roof of the Hokkien Temple. The report's designers picked up the burnished red of the temple's tile roof and repeated it throughout the report, in the endpapers and in Burk Uzzle's flawless photographs of projects for which First Boston arranged financing—expansion of the Atlanta airport, construction of a manufacturing plant in Venezuela, etc. Four interior pages, in which First Boston reviews its year in staggered-length columns of type, are beige. The 38-page report measures 8½ x 11″ (21.6 x 27.9 cm).

Materials and Fabrication: paper: Northwest Quintessence Gloss, 100#; printing: offset lithography, four color with a fifth match color; binding: saddle stitch; cover: liquid laminated.

Client: First Boston Corporation, New York.

Consultant Design: Danne & Blackburn, Inc.: Richard Danne, art director and designer. Photography: Burk Uzzle.

Potlach grows trees and harvests them, converting their wood into lumber, paperboard, and tissues. Headquartered in San Francisco, the company operates throughout the U.S.—in Minnesota, Idaho, Pennsylvania, California, and Arkansas. Twenty-two pages of their 40-page report are a photo and text essay of the company's Arkansas activities, which are, Potlach claims, a microcosm of its operations, from tree planting to lumber and paperboard production. Paul Fusco's photos show the Arkansas forest; the Mississippi River; Potlach employees planting, pulping, cutting, trucking, sorting, and storing; and the Potlach plants at Warren, Prescott, and McGehee, Arkansas. This section on Arkansas is designed to be reprinted separately. Text in the rest of the report is set in two broad columns of type per page with generous white space left at the bottom, beneath the staggered column lengths.

Materials and Fabrication: paper: Northwest Quintessence Gloss 100#; cover: liquid laminated; printing: 4-color offset lithography; binding: saddle stitched; size: 8½ x 11¼" (21.6 x 28.5 cm).

Client: Potlach Corporation, San Francisco, California.

Consultant Design: Danne & Blackburn, Inc.: Richard Danne, art director and designer. Photographer: Paul Fusco. Artist: Marlowe Goodson.

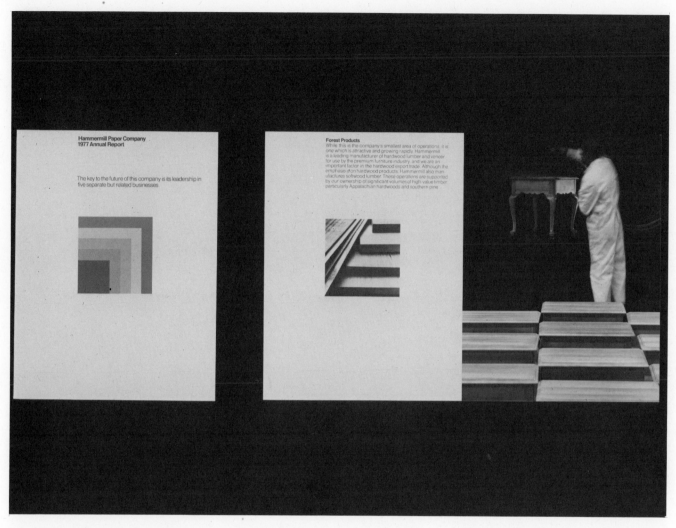

A square in the center of the cover of Hammermill's 1977 corporate report is composed of five overlapping brightly colored squares, representing the company's activities in five related businesses: industrial and packaging papers, fine and printing papers, wholesale distribution, converted paper products (such as envelopes and small paper rolls for business machines), and forest products (hardwood and softwood lumber and veneers). A similarly dimensioned square repeats in the report's first section—a glossy section of color photographs of Hammermill's five separate businesses. In a 3⅛″ (7.9 cm) square opposite each full-page color photo is a photo section of that Hammermill product or service. The report's other two sections are printed on colored stock: beige for the operations review and, logically, gold for the financial review. Size of the 48-page report is 8½ x 11″ (21.6 x 27.9 cm).

Materials and Fabrication: paper: Strathmore (a Hammermill brand) Grandee Duplex and Gold; printing: four-color offset lithography with matte black duotone opposite full-color photos; binding: perfect.

Client: Hammermill Paper Company, Erie, Pennsylvania.

Consultant Design: Danne & Blackburn, Inc.: Richard Danne, art director and designer. Photographer: Cheryl Rossum.

Filmways is a conglomerate corporation composed of holding companies that publish, ensure, entertain (by producing TV series and motion pictures), and manufacture (audio consoles, cartridge tape machines, and automated programming equipment for radio stations). The *Design Review* jurors thought the report's cover gimmicky, looking as if it belonged on an accounting firm's report rather than that of a publishing-entertainment-insurance-manufacturing company. Yet they cited the report's overall quality, its bold, sharp photography, its use of colored horizontal lines as partitions, and its colored stock. Photographs show company employees at work—art directors, film editors, warehouse workers, insurance secretaries—instead of focusing on the glamor that attends their work, on the movie stars and famous authors (Richard Nixon is one) whose work they produce. Report size: 8 x 11" (20.3 x 27.9 cm); 40 pages.

Materials and Fabrication: binding: perfect; printing: five-color offset lithography; cover stock: 100# Quintessence dull with a liquid laminate finish; text stock: 100# Quintessence dull.

Client: Filmways, Inc., Los Angeles, California.

Staff Design: Barbara Shapiro, assistant to the chairman-of-the-board, project coordinator.

Consultant Design: Jonson Pedersen Hinrichs & Shakery: Kit Hinrichs, art director/designer; Tom Tracy, photographer; Jack Weiner, copy writer.

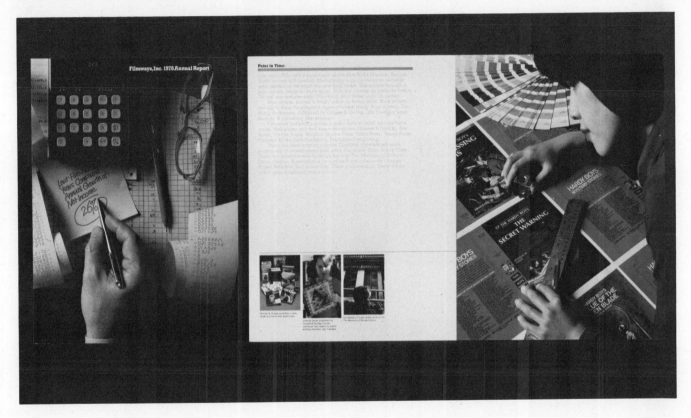

ENVIRONMENT

American Hospital Supply Corporation's Permanent Lobby Exhibit
Smith, Hinchman & Grylls Foyer Light Display
Banners for the Aid Association for Lutherans
Saint Francis Hospital Exterior Signs
Saint Francis Hospital Main Directory
Saint Francis Hospital Interior Signs
Saint Francis Hospital Cancer Center Wall Graphic
Minnesota Zoological Garden Graphic Display System
National Shirt Shops' Interiors
I Natural Cosmetics Shop

Stanley Abercrombie
Stanley Abercrombie is a registered architect and a frequent writer about design. He is editor of a new American supplement to the Italian magazine *Abitare*, and he gave the opening address to the recent Federal Design Assembly in Washington. He has won awards for both design and journalism.

George Nelson
George Nelson, designer, teacher, writer, is president of George Nelson & Company and a partner of Pentagram Design in New York. His awards and honors are numerous, and they started arriving early in his career. After receiving a BA from Yale and attending its School of Fine Arts, he won, in 1932, the Rome Prize in architecture. While in Italy he wrote the first series of articles for U.S. readers on the pioneering work of the modern European architects. These articles led to an editorial job with the *Architectural Forum* where he stayed for nine years, becoming comanaging editor in 1943. During these years he also ran an architectural office with William Hamby, which produced, among other designs, the highly innovative New York townhouse for Sherman Fairchild.

In 1942 Nelson developed the "Grass on Main Street" concept, which became the now-familiar pedestrian mall, and in 1943 he created the Storagewall. His first office furniture, designed in 1946 for Herman Miller, remains an industry standard, and his bubble lamp made possible inexpensive lamps of complex shape.

In 1959 his office designed the U.S. National Exhibit in Moscow, the first major U.S.-Russian cultural exchange exhibit, and it is currently active in products, graphics, office interiors, shops, restaurants, and exhibits.

Nelson is a Fellow of the American Institute of Architects, the Industrial Designers Society of America, the American Academy in Rome, and the Royal Society of Arts.

Robert P. Gersin
Twenty years ago Robert Gersin established Robert P. Gersin Associates, Inc., in New York City. Since then his firm has worked on industrial design projects throughout the world for clients in education, industry, business, and government, garnering from this work more than 200 awards.

Gersin earned a masters degree from Cranbrook Academy of Art and while in the Navy, following his formal education, worked as an industrial design consultant for the Office of Naval Research. Among many current projects the Gersin office is serving as consultant for Expo '82 in Knoxville, Tennessee. Gersin is on the editorial board of *Industrial Design* magazine and the advisory board of Alfred University in New York.

Environment is the most complex category facing annual *Design Review* jurors, just as it is the most perplexing to and least understood by designers. It demands, at its most profound, an understanding of how humans fit into their various environments—offices, factories, retail shops, exhibits, plazas, or cities—and it calls, too, for an understanding of how objects suit these environments, enhancing or degrading them, jarring or blending. Designing a new environment or designing for an existing one requires what one juror called "increasing professionalism," his blanket term for understanding, skill, and care in handling spaces and relationships. "The Beaux Arts building was a unity; everything in it and about it was specified, its dimensions, where it would go, how it would look. We have gotten away from that and are also coming back to it." Today's mobility and change keep more buildings from being designed to house particular interior environments. "An interior designed for one tenant may not suit the next."

Mobility and change are two of the forces that have made environmental design—if not yet a discipline—at least a subject taught in design schools and a skill practiced by a growing number of designers. Population, numbers of people, is another. We are buffeted by the compression and distortion of too many people in too little space. As 1978 began, demographers pegged U.S. population at 217.7 million, an increase of 1.7 million since early 1977 and of 14.5 million (half a million more than the entire population of Australia) since the 1970 census. Since 1960 the U.S. has gained 38.4 million people and in the 28 years since 1950 an awesome 66.3 million. In the past 28 years the nation has added more people than were here in 1890, more than there are in Great Britain, or France, or West Germany . . . or Mexico. Since 1920 the U.S. population has doubled. There are now 57.5 of us for every square mile of U.S. land, but of course we crowd into cities where this statistic may seem a relief. Cities shelter 157.5 million of us, versus about 57 million in nonmetropolitan areas; and though since 1970 there has been a

slow trickle of migration away from cities, metropolitan areas added, in this decade's first seven years, 7 million people; nonmetropolitan areas gained 4 million.

The numbers do not stop with people. More people wear more jeans, buy more dishwashers, dig more raw materials to feed more factories, cut more trees so more offices can distribute more paper, build more roads and trucks to bring everything together and take it away, until we are as compressed by our trash heaps, factories, and department stores as we are by our neighbor's blaring television, ringing phone, or backyard jungle gym.

If this crowding seems obvious, the effect it is having on designers is not. The profession is moving slowly into uncharted areas where an object must be considered in relation to its surroundings, and where increasingly a designer is being asked to design these surroundings—settings and spaces that will make workers more productive, shoppers more likely to buy, visitors more apt to absorb and retain information.

Perhaps because of this category's complexity and relative newness to design, Environments drew only 36 entries for review by the jury, but of these the jurors—Stanley Abercrombie, Robert P. Gersin, and George Nelson—chose more than a quarter (10) for display here. The jury's approach was relaxed, contemplative, as they considered which retail shops, exhibits, signs (environmental graphics), office and factory interiors, plazas, even festival designs, to select. They wanted environments that work, that solve their particular problems tastefully and definitely; but they wanted, too, environments that could be illustrated well in these pages. They hesitated over the foyer lighting display (page 164), designed by Smith, Hinchman & Grylls, the Detroit-based architectural and engineering firm, because the lighting, a play of shaped colored lights, might not show up well in black and white. Their reluctance to do it an injustice gave way to their regard for the effect the display gives the lobby, setting it apart from others, and for the fact that its designers worked with abstractions.

Another lobby exhibit (page 160) is far from abstract. American Hospital Supply Corporation wanted a permanent display touting the company and its role in American medicine, and though the jurors liked parts better than the whole, they felt the exhibit prepares a visitor for the corporation beyond it, that the exhibit, if not a spectacularly successful environment on its own, shapes the way a visitor will perceive the environment he's moving into.

While one juror noted that well-designed display devices are becoming more widely available, making it possible for a designer to put together a good exhibit with stock components, another juror pointed out that trade shows in particular, where these stock components are used, show little design improvement. "Trade shows are a one-shot operation," he gave as a reason, "they only have to lure people once." Their voice is strident, their environment brash.

Unlike trade shows retail stores are changing, beginning to recognize new design rules. Shopping, of course, no longer takes place on Main Street. It thrives in a drive-in complex anchored by two department stores. Far from enthusiastic about shopping centers and malls, the jurors maintained that environmental design is helping retailers. It is possible, of course, that a *lack* of design may be a sort of shopping lure. Equating no-frills display with low prices, people may be drawn to a K-Mart, noted one juror. But, nonetheless, shopping centers have an increasing number of objects competing for space, and design can help them gain attention. Malls must first lure shoppers; then each retailer there must compete with his peers for attention, like exhibitors at a trade show. But the comparison goes no further. "A retailer in a shopping mall has a long-term responsibility," said one juror. "He must not only lure shoppers, he must also lure the same shoppers repeatedly, the people in the neighborhood who form the bulk of his clientele." A retail store that lures shoppers repeatedly and makes them want to linger once in it is National Shirt Shops clothing store (page 176). National Shirts is a chain, whose shops appear mostly in shopping malls, and which now offers shoppers an appealing, ordered, relaxed atmosphere for shopping.

A colorful innovation in retail environments is the I Natural Cosmetic Shop (page 178). It borrows retail techniques from the past, putting them to a new use in a shop which looks contemporary but is really borrowing an appealing, comfortable atmosphere from the past.

Combining elements from trade exhibits and retail shops is the Hertz Corporation's reservations desk, or as they call it "customer area." These areas are designed to specifications that cover everything from lighting to the width of the attendant's hair band. Though the jurors did not want the customer area singled out for display here, they felt it represents a neat organization of information and procedure. Your reservations form is waiting, identified by your name in bold magic marker inking, so you can spot it in a rack behind the desk as you move toward it. Everything seems to be simplified, organized, ready to go, as if Hertz understands your rush and is ready to help. The display area is modest and the lighting is sensational, argued one juror. But another maintained with conviction that "It's sort of cheap and nasty. And I don't like the look of materials." The jurors agreed that though the area's design is not impressive, it does work: it does suit a special situation.

Perhaps even less well understood than the environments of retail stores and exhibits are office environments. "Numbers of people are part of the office game," said one juror. "Offices now house the greatest proportion of working people." And another juror added, "We'll probably see more of the open office because of the efficiency it offers in getting more people into the same area." Even so, "we understand little of how people work in offices," said a juror. "We may say person 321 does concentrated work so we immediately think he needs an enclosure. But perhaps he works best in the midst of bustle. We'd have to find out whether he's Virgo or Capricorn, what his nuances are, then design for how he works best." But offices by nature are impersonal, or at most, personal only in their conformity to some mythical personal norm. "It is hard to imagine a personal office," the juror went on. "It's the reason people fill

their work spaces with personal treasures—family pictures, water polo trophies from their sophomore year. These personal objects are lifelines."

What many may need for effective work is a view. "I have the best work space I've ever had," explained one juror. "I've placed my desk by a window so I can look out over trees in the park. I can look up from my work, focus on the park, and look back refreshed." By contrast, a Wall Street executive, whose office high in a financial district tower looks out over the dots of tugs and freighters in New York harbor far below, dismissed a visitor's comment on the view by saying, "After a while it becomes wallpaper." Not everything is understood about how people work.

If open offices make paper processing easier than cubicle offices do, open plans produce other problems. Open offices highlight, perhaps even exaggerate, problems of privacy and personal territory. White noise is being tried to mask neighboring workers' phone conversations. But for the most part open office problems remain unsolved while design effort focuses on technology, on refining open office furnishings. Wiring is being tucked unobtrusively into these furnishings and their work surfaces being given special lighting (partly in an effort to save energy). A few attempts even bring humor and humanity to these furnishings, but the technological refinement has not, as one juror said, produced "parallel development in visual aspects of office design. The move from modernism to postmodernism, though talked about a lot, hasn't showed up in new things." No office environments appear here.

Nor do manufacturing environments, though the jurors liked what Westinghouse did with one of its manufacturing facilities. Westinghouse designers with consultant architects Hobart Betts Associates renovated the plant, using light color on the walls, bright colors on machines and equipment, replacing mercury vapor lighting with sodium vapor. The space is unquestionably less gloomy, and plaque signs with clear, color-coded graphics placed on girders and walls make the plant less confusing. Perhaps most important the redesigned environment is more pleasant for workers and visitors, perhaps even increases the plant's efficiency and productivity. But in light of what is being done to other manufac-

turing plants, the jury considered it nonextraordinary. It is merely an example of what manufacturing environments are moving towards, and while the jury wanted to encourage the direction, it did not want to single Westinghouse out. Factories, like offices, still have much to tell us about how people work in them. Music and story-readers may occasionally relieve assembly-line tedium but are, of course, only a palliative.

Of all the environmental arts, environmental graphics has come the farthest, is the best understood, and has the most practitioners. "Most architectural offices now have graphic departments and recognize the need for consultants if they don't," said a juror. Environmental graphics has become widespread enough to evolve its own cliches. It has come to a point where, as one juror noted, probably "sign systems themselves are already a cliche." And certainly an overreliance on Helvetica typeface is another. Environmental graphics, of course, goes beyond signs to the other graphic accoutrements that shape and enhance a given environment while suiting it exactly.

Again our mobility and numbers, of people and institutions, give environmental graphic design an urgency. "More people are transient, passing through large spaces in mammoth airports, hospitals, hotels, and office buildings where they need guidance and reassurance.

The jury singled out six examples of environmental graphics for inclusion here. Most striking are the massed architectural structures that are the graphic displays at the Minnesota Zoo (page 174) and the exterior signs at Saint Francis Hospital in Tulsa (page 168). In size and form they complement splendidly the buildings they direct people to.

Inside sign systems are represented by the interior signs of Saint Francis Hospital (page 170). Most of these signs are lighted from within. This lighting focuses their messages, helping them stand out in the institution's seeming confusion and the glare of white lighting that washes its corridors. Interiorly lighted signs appear more frequently today, perhaps because they focus, and thus save, energy. But they are especially useful, said one juror, "where a designer can't control the other lighting around a sign."

Absent from this year's *Design Review* are plazas and courtyards, though the jurors noted gains in knowledge of how courts are used, how they should be designed, and what makes them enhance an adjacent environment. Courts need to be woven into a city's fabric, the jurors agreed. Without plazas a walk through a city becomes "like walking through a series of corridors without encountering a room." Increasingly courts are being enclosed so they can bloom all year, providing gardens in the midst of winter-gripped northern cities. Most often these courts nestle up to buildings or are actually enclosed by a building's structure, and there is in general more attention being paid to the relationship of the building to these courts and, by extension, to the street beyond. These relationships are becoming "richer, warmer, more varied."

But despite environmental design's complexity, examples of the art, seen here, have like the objects and graphics seen in this book's other sections, a common simplicity and economy of expression. They have, one juror said, "an imaginative use of the simple and the inexpensive. The reason for this is today's taste and economics. Items that are nonostentatious and non-wasteful are more appealing." Said another more directly: "The controlling force of what's happening today is dollars and cents. People ask: 'What is it going to cost?' Still, despite this concern, there is a consciousness of the effect places have on people." Whatever the constraints—ignorance, budgets, sloth, inertia, greed, numbers—the effect of environments, one can see here, are often benign, promising enough to make us want to go further.

Renovation of Westinghouse manufacturing facility. Client: Westinghouse Electric Corp., Pittsburgh, Pennsylvania; staff design: Richard Klein, Steven Quick; consultant design: Hobart Betts Associates, Architects, New York.

Two components of American Hospital Supply Corp.'s headquarters lobby exhibit caught the jurors' attention. Outstanding, they felt, and what they wanted to single out for inclusion here, are the "biomedical man," the manikin put together with several of the company's prosthetic devices, and the mirror-backed case display of the company's single-use items—surgical scissors, tweezers, stethescopes, gloves, droppers, etc.—set up in striking, repetitive rows on glass shelves. The entire exhibit is meant to show anyone passing through the lobby (including community and school groups) just what American Hospital Supply does and what the corpora-

tion's role is in health care development. Displays are housed in glass cases with laminate bases and lightbox tops; case dimensions are 27 x 54 x 84" (68.6 x 137.2 x 213.4 cm). Central exhibit element is what the designers call a "timeline," a lightbox flowing through the exhibit 30" (76.2 cm) off the floor. It holds transparencies, lighted from below, identified by silk-screened graphics, offering glimpses of the history of medicine. Elsewhere the exhibit relies on push-button display video tapes and a demonstration of the computerized network American Hospital Supply uses to distribute its 113,000 products.

"Prosthetic" Man

"Prosthetic" man—not a citizen of some far future day, but a representation of a person who could live today with the aid of artificial body-parts, some of them engineering feats of the 1970s.

The prosthetic devices exhibited here are a few of those manufactured and distributed by units of American Hospital Supply Corporation.

Intraocular lens. This lens, made of clear acrylic and polypropylene materials, can be surgically implanted in the eye to replace a natural lens that has become clouded by a cataract. In one recent year in the United States, some 100;000 intraocular lenses were implanted. The lens is manufactured by the Heyer-Schulte Medical Optics Center.

Dentures. New plastic materials and precise fitting techniques have improved the comfort, durability and appearance of artificial teeth today. These are distributed by the Denticon division.

Hearing aid. This and other types of hearing aids are distributed by the HC Electronics, Inc., subsidiary.

Chin prosthesis. This artificial chin, made of silicone, is implanted under the skin in cosmetic or reconstructive surgery. The body's own tissue affixes itself to the prosthetic chin, growing through the small holes around the edge. During the past two decades, development of many new implantable prostheses has been made possible by the use of silicone, a material less likely than some to be rejected by the body. This and other implantable devices are manufactured by the Heyer-Schulte Corporation subsidiary.

Mechanical heart valve. Manufactured by the Edwards Laboratories division, valves of this sort were introduced in 1960, after research by Dr. Albert Starr, in collaboration with Edwards. This one is used to replace a malfunctioning aortic valve in the heart. The ball in the valve moves up and down in a metal "cage", regulating the blood flow.

Pacemaker. An artificial pacemaker, implanted in the chest, sends small electrical pulses to the faltering heart, helping it maintain a regular beat. This pacemaker is "programmable" even after it is implanted, to allow adjustment in the pulse rate. It was developed and is manufactured by the Edwards Pacemaker Systems division.

Artificial arm. To replace an arm amputated below the elbow. Hosmer-Dorrance manufactures 21 varieties of this hook-like hand, to aid in restoring lost manual abilities.

Artificial arm. Manufactured by the Hosmer-Dorrance Corporation, to replace an arm that has been amputated above the elbow.

Arterial graft. This graft, made of Dacron, is used to replace a section of artery that has been damaged—or has become weakened, presenting the potentially fatal danger that it could burst. Grafts for various blood vessels are distributed by the V. Mueller division.

Artificial leg. A Hosmer-Dorrance prosthesis to replace a leg that has been amputated above the knee.

Orthosis. This device is used to support and help maintain the function of a leg in which joints or muscles have been weakened by injury or disease. It is manufactured by Hosmer-Dorrance.

Materials and Fabrication: cases are ¼″ (6.4mm) clear plate glass with clear anodized aluminum trim. Bases are laminates with Plexiglas overlays. The timeline case has a substructure of cold rolled steel to which laminate and Plexiglas are secured; upper portion is ½″ (12.7 mm) polished plate glass trimmed in clear anodized aluminum extrusions.

Client: American Hospital Supply Corporation, Evanston, Illinois.

Manufacturer: Proto Impressions; Robert Peterson Design.

Client Design: Jeff Rich, director, communication services; Kerry Bierman, manager, de-

sign; Tom Salisbury, senior designer; David Bates, designer; Judy Benoit, assistant designer; Phil Smith, manager, editorial services; Nancy Hobor, manager, planning, research.

Consultant Design: Norman Perman Design: Norman Perman, owner; Elaine Kobold, designer. Proto Impressions: Ken Hopkins, owner; Gary Grube, partner; Kyoji Nakano, designer. Harling, Winer & Shiromani: Mel Winer, partner. Joseph Sterling Photography: Joseph Sterling, owner. Cynthia Anderson Design: Cynthia Anderson, designer. Teletronics: Hal Rein, director/producer. K&S Photographics: John Stein.

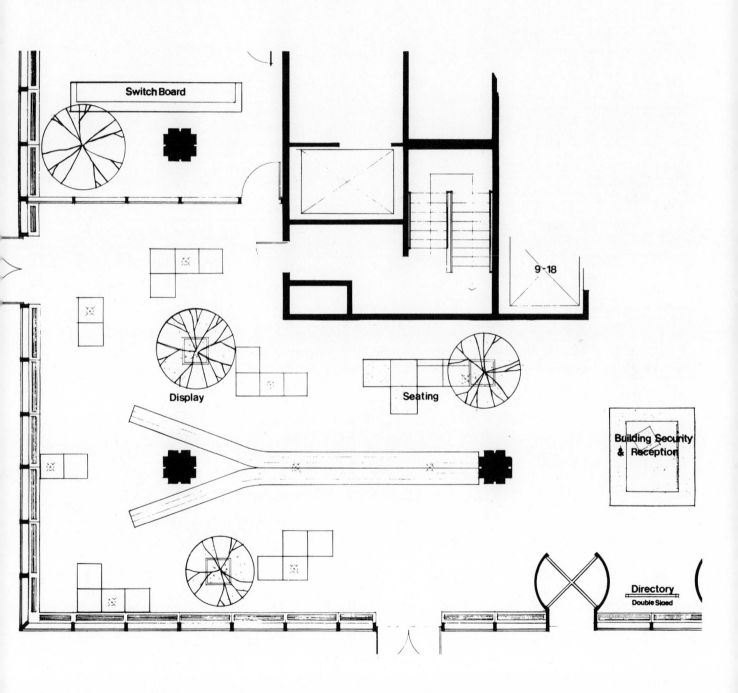

Using a series of track-mounted Kleig lights that move back and forth and pivot like dancers, Smith, Hinchman & Grylls Associates, Inc., the Detroit-based architectural and engineering firm, created a light display for their headquarters building lobby. Colors and shapes thrown by the lights onto the walls, and onto the floor 15' (4.6 m) below, vary almost infinitely. Shapes are controlled by four framing shutters working behind mask inserts and built-in frames. Colors are created by gel filters. Variable focus determines the images' sharpness, and dimmers control the light intensity. Though the jurors were not overwhelmed by the light display ("Very pretty," they called it), they noted that it is a rare example of a professional designer trying to deal with abstract information.

Materials and Fabrication: theatrical Kleig lights in a ceiling-mounted light track.

Client: Smith, Hinchman & Grylls Associates, Inc., Detroit, Michigan.

Staff Design: Steve Stannard, illumination engineer.

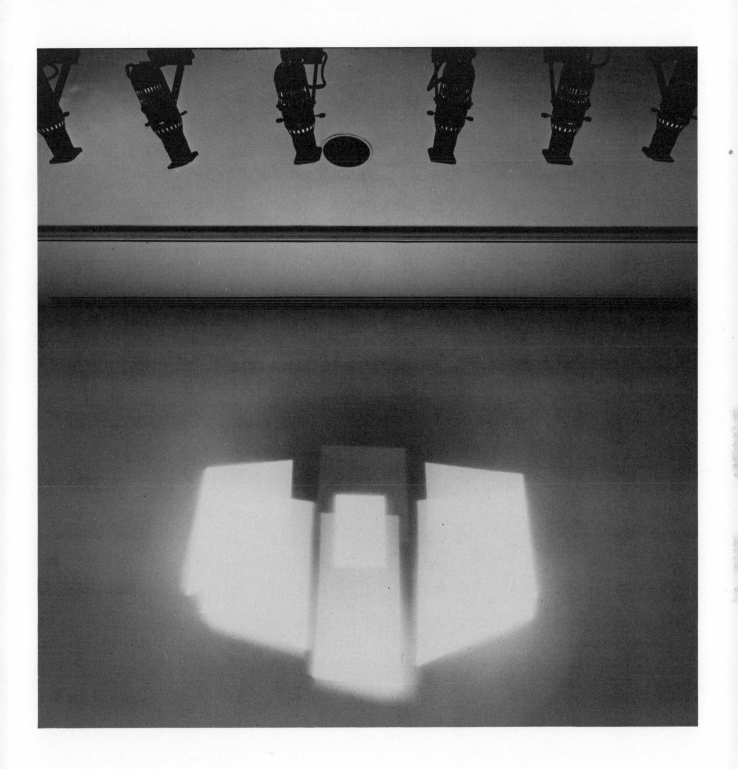

In all the Aid Association for Lutherans commissioned 35 banners, 25 of which measured an impressive 9 x 5′ (2.7 x 1.5m). Used both as decorative art and as informative landmarks to guide people through an open plan landscape, the banners are a blend of techniques (applique, trapunto, batik, weaving, and ink on canvas) and materials (cotton, felt, canvas, nylon), and a cornucopia of natural subjects—animal (sheep, moose, frogs, fish, birds, beetles, butterflies, etc.) and vegetable (apples, corn, fruit trees, a flower garden, etc.). Six banners by Norman LaLiberte, measuring 11½ x 3′ 4½″ (3.4 x 1m) hang in the Association's cafeteria and show subjects and scenes such as the art of cooking, the art of eating, hunting for food, the sun and the ox.

Elsewhere four giant [21½ x 4′ (6.4 x 1.2 m)] club banners by Anders Holmquist have geometric designs based on wrought iron patterns. The only instructions given the 14 artist-craftsmen who designed the banners pertained to sizes, which were predetermined, and clarity of communication: designers were asked to stick to subjects easily understood and simply described, perferably in a word or two.

Materials and Fabrication: nylon, cotton, canvas felt; applique, batik, weaving, trapunto, ink drawing.

Client: Aid Association for Lutherans, Appleton, Wisconsin.

Consultant Design: George Nelson & Company. Individual artists: Laura Adasko, Evelyn Anselevicius, Carolyn Bell, Margaret Cusack, Kristina Friberg, Jenet Hoffman, Anders Holmquist, Howard Koslow, Norman LaLiberte, Karen Lawrence, Jeanne McDonagh, Dina Schwartz, David Stone, Yoshi.

Depending on where they're positioned on the campus of Saint Francis Hospital in Tulsa, the signs which carry traffic commands and information about destinations on the grounds are either four- or two-sided. Four-sided signs, square pylons, are constructed of either 3' or 1'6'' (91.4 or 45.7 cm) panels. Thin, slab-like plinth signs have information on only two sides and are made up of interchangeable panels 4' (121.9 cm), 3' (91.4 cm), or 1'6'' (45.7 cm) square. Largest panels, carrying the largest signs and blocks of type, are, obviously, used where traffic flows fastest, such as at the hospital's entrance; smallest panels are used where traffic is slowest. Information is positioned according to its importance and that importance is reinforced by color. The emergency entrance, for instance, is announced at the top of a pylon in red letters. The designers made an effort to limit the amount of information carried on any one sign, trying to feed information to passersby in small, easily absorbed amounts. So the directions are succinct—Main Entrance, for instance, followed by an arrow pointing in the direction to be taken.

All the hospital's exterior signs are lighted from within by cold white, high-output fluorescent fixtures. Letters are cut into the aluminum surface of the sign panels, covered on the inside by an acrylic diffuser and then lighted from within.

Materials and Fabrication: Modular panels are of ⅛'' (3.2 mm) sheet aluminum, which are bent on a brace and welded at the corners. Copy is precision cut into these panels with a Gorton Pantomill. Panels are then anodized a dark bronze duronodic.

Client: Saint Francis Hospital, Tulsa, Oklahoma.

Manufacturer: Claude Neon Federal, Tulsa, Oklahoma.

Staff Design: Sister Mary Blandine, administrator; Robert W. Degen, assistant administrator in charge of project; Lloyd Verret, assistant administrator; Dr. Robert Tompkins, medical director.

Consultant design: Robert P. Gersin Associates, Inc.; Robert P. Gersin, program director; Ingrid G. Caruso, project director; Kenneth Cooke, graphic design; James Goldschmidt, product design.

The triangular ends of Saint Francis Hospital's main building directory angle in, delineating a semi-private space for someone using the directory. Besides that they offer an exterior surface used by the designers to announce the directory's presence. "Directory" is proclaimed in letters nearly a foot high on either end's exterior, boldly enough to keep visitors from wandering aimlessly looking for assistance. Yet even without graphics the directory would be hard to miss or mistake. Measuring 22 x 5 x 4' (7.5 x 1.5 x 1.2 m), "it's a very important object that you can't miss, yet it's not protrusively fussy," noted one juror.

Type and diagrams on the directory surface establish the hospital's four main areas and the color that represents each. These colors carry through on the hospital's interior signs (see page 170). The directory also displays floor diagrams, announces visitors' hours, and lists departments. A section on each triangular endpiece is set aside for variable information, such as the dates of hospital lectures or symposiums.

Behind the directory's colored areas, information and graphics are set on film negatives and roscolene gels in sandwich panels with faceplates of ¼" (6.4 mm) bronze polycarbonite. Cold cathode white tubes set on 6" (15.2 cm) centers light the sign from within.

Materials and Fabrication: triangular end sections are ⅛" (3.2 mm) thick aluminum plate, finished with a thermal setting acrylic. Center portion of the directory is an aluminum extrusion inset with a graphic polycarbonite sandwich panel.

Client: Saint Francis Hospital, Tulsa, Oklahoma.

Manufacturer: Claude Neon Federal, Tulsa, Oklahoma.

Staff Design: Sister Mary Blandine, administrator; Robert W. Degen, assistant administrator in charge of program; Lloyd Verret, assistant administrator; A. Robert G. Tomkins, medical director.

Consultant Design: Robert P. Gersin Associates, Inc.: Robert P. Gersin, program director; Ingrid Caruso, project director; Georgina Leaf, graphic design; Gabrielle Crettol, project design.

Once inside Saint Francis Hospital and beyond the main directory (see page 169), a visitor or staff member can find his way logically through the building's complex maze of corridors, levels, wings, departments, offices, wards, laboratories, operating rooms, and nursing stations. The sign system designed by Robert P. Gersin Associates to guide hospital travelers is simple, keeping, like a politician in an election year, each individual message uncomplicated. In this case the message is reinforced by using plenty of signs so hospital visitors who may be distraught or distressed have little chance of losing their way. At the main directory visitors see that a color represents each of the hospital's four major wings. These colors appear consistently on all signs throughout the hospital, first in a series of internally illuminated signs—wall, floor, or even ceiling mounted, leading visitors and staff to the area they seek. Each signface has no more than one directional arrow. And all elevator lobbies have additional directories. For the very specific information needed to lead visitors to, say, a particular patient room, office, or department, plaque signs take over. These have colored squares in the upper left-hand corner, identifying the hospital area one is in. To keep the identity of staff facilities and visitor paths separate, staff signs have a white ground with dark letters. Visitor signs are dark with white copy. In addition a supporting, complementary sign system consists of changeable cards in a frame of the same ABS polymer used in the hospital's other nonilluminated signs. The variable information cards in these latter signs are styrene with silk-screened graphics, and each has blank areas where changing information can be penned in.

Illuminated signs are based on a 15" (38.1 cm) square module that is 4" (10.2 cm) deep. Floor signs stand 60" (152.4 cm) high. Nonilluminated signs are based on a 7½ x 10½ x ⅝" (19.1 x 26.7 x 1.6 cm) unit.

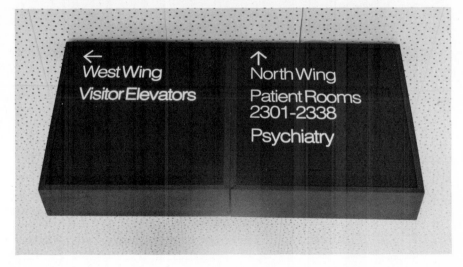

Materials and Fabrication: illuminated signs: extruded aluminum (6063 alloy) finished with a thermal setting acrylic; graphic sandwich panel: ½″ (12.7 mm) bronze polycarbonite; into the sandwich slide a film negative with roscolene gels and an ⅛″ (3.2 mm) thick translucent acrylic diffuser; these back the signs' letters, coloring the light thrown by each sign's cool white fluorescent tube.

Nonilluminated signs are injection molded ABS with face plates of ⅛″ (3.2 mm) non-glare acrylic.

Client: Saint Francis Hospital, Tulsa, Oklahoma.

Manufacturer: Claude Neon Federal, Tulsa, Oklahoma.

Staff Design: Sister Mary Blandine, administrator; Robert W. Degen, assistant administrator in charge of project; Lloyd Verret, assistant administrator; Dr. Robert G. Tompkins, medical director.

Consultant Design: Robert P. Gersin Associates, Inc.: Robert P. Gersin, program director; Ingrid Caruso, project director; Kenneth Cooke, Georgina Leaf, Barbara Daley, Gabrielle Crettol, James Goldschmidt, project design.

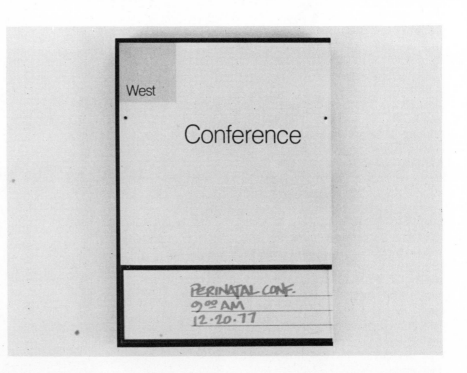

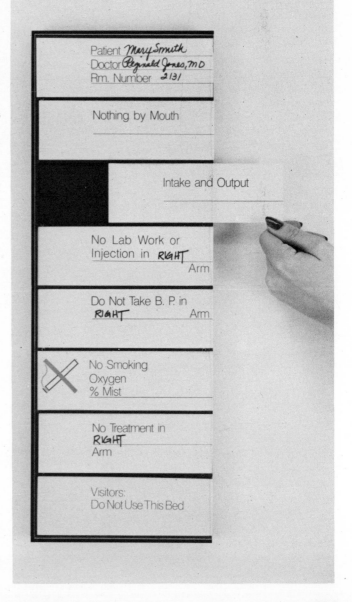

Velcro-tipped dowels placed in a 2' (61 cm) grid on a wall become hooks for Velcro pads mounted on the backs of brightly painted modular shapes. By moving the shapes from peg to peg anyone can rearrange the wall graphic, temporarily changing the entire environment, of which the graphic on a 19 x 9' (6 x 2.8 m) wall is the dominant element. The environment is a skylit court between two waiting areas of Saint Francis Hospital's Natalie Warren Bryant Cancer Center, where it provides a gentle, quiet outlet for tensions of both staff and patients. The movable modular shapes are flowers, clouds, and waves, and because the grid on which they must be placed imposes a sort of loose discipline, any rearrangement becomes a pleasant wall graphic.

Materials and Fabrication: movable modules are of ¾" (19.1 mm) plywood cut to shape. Graphics are silkscreened or frisket on a colored laquer base with a clear laquer finish. Pegs of 2½" (6.4 cm) diameter wood dowels are attached to existing sheet rock wall.

Design Control Drawing

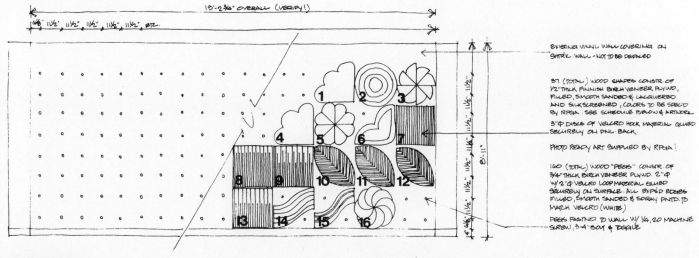

Elevation
SCALE: 1/2" = 1'-0"

Handwritten notes (right side):

EXISTING VINYL WALL COVERING ON SHTRK. WALL · NOT TO BE DEFACED

37 (TOTAL) WOOD SHAPES CONSTR. OF 1/2" THICK FINNISH BIRCH VENEER PLYWD. FILLED, SMOOTH SANDED & LACQUERED AND SILKSCREENED ; COLORS TO BE SPEC'D BY RPGA · SEE SCHEDULE BELOW & ARTWORK

3"⌀ DISCS OF VELCRO HOOK MATERIAL GLUED SECURELY ON PNL. BACK.

PHOTO READY ART SUPPLIED BY RPGA!

160 (TOTAL) WOOD "PEGS" CONSTR. OF 3/4" THICK BIRCH VENEER PLYWD. 2"⌀ W/ 2"⌀ VELCRO LOOP MATERIAL GLUED SECURELY ON SURFACE. ALL EXP'SD EDGES FILLED, SMOOTH SANDED & SPRAY PNTD. TO MATCH VELCRO (WHITE)

PEGS FASTND TO WALL W/ 1/4, 20 MACHINE SCREW, 3-4" BOLT & TOGGLE

Detail
SCALE: FULL SIZE

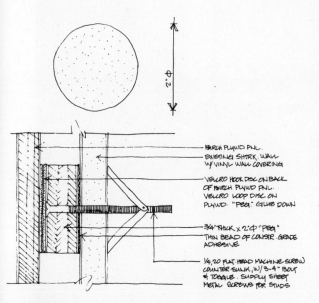

Handwritten notes (Detail):

BIRCH PLYWD PNL.

EXISTING SHTRK. WALL W/ VINYL WALL COVERING

VELCRO HOOK DISC ON BACK OF BIRCH PLYWD PNL. VELCRO LOOP DISC ON PLYWD. "PEG" GLUE DOWN

3/4" THICK x 2"⌀ "PEG" THIN BEAD OF CONSTR. GRADE ADHESIVE

1/4, 20 FLAT HEAD MACHINE SCREW COUNTER SUNK, W/ 3-4" BOLT & TOGGLE · SUPPLY SHEET METAL SCREWS FOR STUDS

Schedule

PIECE NO.	DESCRIPTION	NO. REPEATS	NO. COLORS	NOTES
1	FLOWER/CLOUD	2	1	NO'S 1 & 4 SAME SHAPE
2	SUNBURST	1	2	
3	YELLOW FLOWER	1	2	
4	FLOWER/CLOUD	3	1	SAME SHAPE AS NO.1
5	BLUE FLOWER	1	2	
6	MAGENTA PETAL	4	2	
7	GRASS/RAIN	2	2	NO'S 7, 8, 9 & 13 SAME SHAPE, 4 SCREENS
8	GRASS/RAIN	1	2	
9	GRASS/RAIN	3	2	
10	LEAF A	2	2	SAME SHAPE, SAME SCREEN, 3 COLOR COMBINATIONS NO'S 10, 11, 12
11	LEAF B	2	2	
12	LEAF C	2	2	
13	GRASS/RAIN	2	2	
14	EARTH/WATER	6	2	
15	AIR/WIND	4	2	
16	MAGENTA FLOWER	1	2	

Client: Saint Francis Hospital, Natalie Warren Bryant Cancer Center, Tulsa, Oklahoma.

Manufacturer: CDI Industries, New York.

Client Design: Sister Mary Blandine, hospital administrator; Robert W. Degen, assistant administrator in charge of project; Lloyd J. Verret, assistant administrator; Dr. Robert G. Tomkins, medical director.

Consultant Design: Robert P. Gersin Associates, Inc.: Robert P. Gersin, program director, graphic design; Ingrid Caruso, project director; Barbara Daley, graphic design.

Built on 480 acres of rolling farm woodland, interspersed with small lakes that are 30 minutes from the centers of Minneapolis and St. Paul, the Minnesota Zoological Garden is the first northern climate zoo that is open all year. About two-thirds of the main building is underground, offering below-grade circulation to the major zoo exhibits, which as far as possible present animals in their native habitats. The exterior signs, identifying and providing information about different animals and habitats, were also designed by the architects. They rely on the signs to provide strong forms on the zoo's entrance way, and as a result, the signs have become pieces of architectural sculpture, towering as high as 14' (4.2 m), distributing color, information, and a sense of place. Information is presented either by silkscreening it directly on enamel-covered aluminum panels or by enclosing ciba print enlargements in acrylic double panels. These photo panels of animals and their natural habitats are mounted on multifaceted pylons fastened to 6" (15.2 cm) concrete bases. Closed circuit TV screens, placed in the frames, can show the animal being discussed in its barn or den.

In form these signs are architectural, meant to complement the buildings around them, and one juror pointed out that the signs' form is paramount, that they "could look quite pleasant in whatever setting (at the zoo) they are placed." Another juror, though agreeing, found the relationship of the signs' form and the material they carry self-conscious.

Materials and Fabrication: frame 1 x 1" (2.5 x 2.5 cm) .100 aluminum tubing painted with a baked epoxy enamel. Infill panels are frames with .125 aluminum panels with baked-on enamel on which graphics are silkscreened. Acrylic double panels hold ciba print enlargements.

Client: Minnesota Zoological Board, Apple Valley, Minnesota.

Manufacturer: Colite Industries, Inc., West Columbia, South Carolina.

Client Staff: Bev Rongren, Education and Interpretive Services, coordinator.

Consultant Design: InterDesign Inc.: Peter Seitz, project director and designer; Sanford Stein, project designer; Hideki Yamamoto, staff designer. Special art consultant: Concept, Inc.

National Shirts has 280 stores, mostly in shopping malls. A typical store is long and narrow, about 30 x 100' (9 x 30m) and its customers are predominantly men, who buy the shop's middle-priced men's wear. To heighten the stores' appeal to these customers, the designers used deep, rich colors, mostly reds and browns, throughout the stores and kept the materials—natural tile, fabric, and wood—roughly textured. To reduce the sense of narrow space they set up the stores on a diagonal, creating three levels, each separated by a couple of steps. These levels help divide types of merchandise and make that merchandise more visible from different spots within the store. "It's a real accomplishment of organization," said a juror, "to have the store look so clean and clear, while holding so many bits and pieces of shirts. There is something very consistent about the angles of the ceilings and walls." On the first level are scattered racks of hanging clothing and formal displays—mannequins, standing on cubes, wearing whatever clothes are in season. The second level has a specially designed wrap-cash desk, a small seating area (also specially designed) with textured brown fabric on the comfortable upholstered seats, merchandise on shelves and racks and two freestanding, mirror-wrapped dressing rooms. On the third level are more racks and shelves of clothes. Lighting throughout is dramatic. Pools of bright light beamed from spotlights mounted on top of the dressing rooms and recessed in the ceiling surrounding clothing displays, and light bars, cantilevered from the wall above

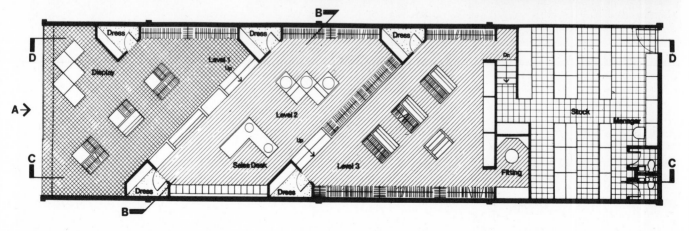

shelves and racks, wash the merchandise there. Spots also wash the stair risers between floor levels, and a table lamp throws an intimate pool of light onto the wrap desk.

To make seasonal display changes easy, the designers specified modular shelving and hang rods. Graphics on transilluminated film strips on the cantilevered light bars are easily changed and promotional posters fit onto floor displays. Throughout an image is created by the use of light, texture, and color and by grouping

clothing, putting sportswear, outerwear, and accessories, say, together rather than isolating each.

Sales in National Shirt's redesigned shops have increased, they claim, an amazing 60 percent.

Materials and Fabrication: acoustical ceiling tile; prefinished red oak flooring; kelly green suede vinyl wall covering; umber quarry tile; mirrors.

Client: National Shirt Shops, New York.

Staff Design: Edward Marks, president; Steven Marks, assistant to the president; Daniel Mark-

owitz, vice-president and general merchandising manager; Robert Stone, vice-president and director of real estate.

Consultant Design: Robert P. Gersin Associates, Inc.; Louis Nelson, Robert P. Gersin, program directors; Lee H. Stout, project director; Judith Lane, Gabrielle Crettol, interior design.

The I Natural retail cosmetics shop brings the retailing methods of the country store and the neighborhood tavern to cosmetics sales. Instead of buying a bag of crackers from a barrel or a pint of beer from a keg, the I Natural customer buys a moisturizer, shampoo, or conditioner from 5-gallon (19-liter) glass jars. Pressing a lever behind the counter much as a bar maid or soda jerk would, the cosmetics store clerk fills a container for the customer [containers come in 4-, 6-, 8-, and 16-ounce (.11-, .17-, .23-, and .46-kilogram) sizes]. Since the shop is tiny [15 x 17' (4.6 x 5 m)], it can hold more cosmetics in bulk than it could by stocking innumerable prepackaged small jars. Except for green stool covers and green benches, forming a booth by the front window, the shop is white, white ceramic tiles on the floor, white ceilings and walls, white countertops, all with stainless steel trim. The cosmetics provide the color, displayed in glass jars in glass-shelved wall cabinets and on the countertop.

The jurors praised the shop's intimacy. They found its approach a refreshing change from most cosmetics shops and counters. "It has a fresh, spring, dreamlike quality," said one. Commented another, "It looks old fashioned without having any period elements in it."

Materials and Fabrication: ceramic tile, painted wood, plastic laminate, and stainless steel.

Client: I Natural Cosmetics, New York.

Staff Design: Lois Muller, president.

Consultant Design: Walker/Group, Inc.: Kenneth H. Walker, president; Robert L. Turner, project director, director graphic and industrial design; Frank Koester, project designer.

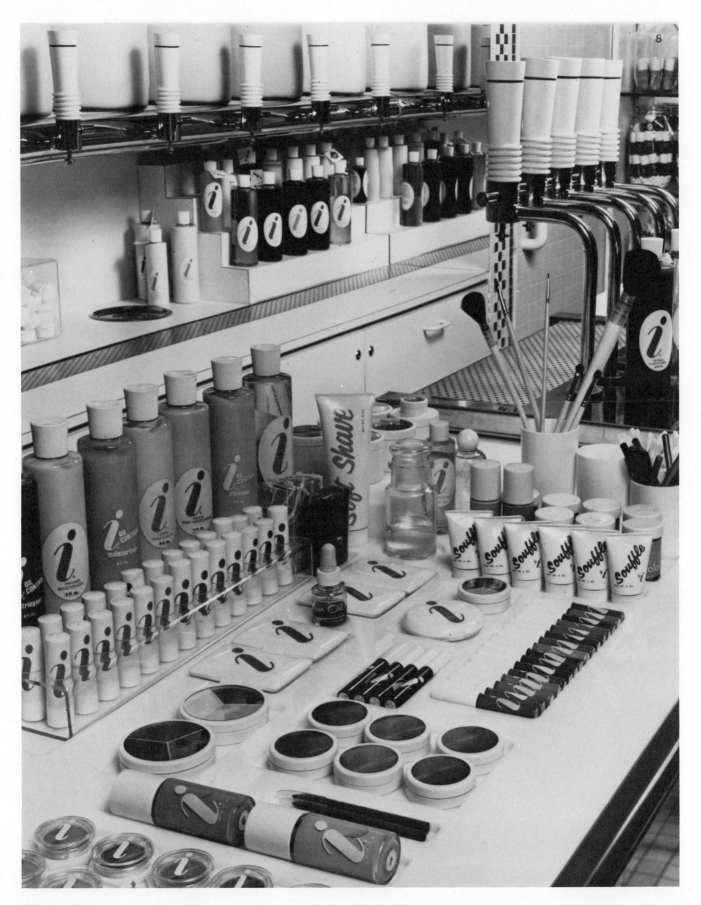

CREDITS

Photographers are listed below according to the pages on which their work appears:

63: Wine Image Photography

70, second from bottom: Stan Ries

70, bottom: Guiseppe Pino

74-75: Michael Pateman

127: Ernest Pappas

128-130, 133-134: Bernard Askienazy

178-181: Victoria Lefcourt

DIRECTORY OF DESIGNERS

Consumer Products

Marc Harrison Associates   50
568 Bristol Ferry Rd.
Portsmouth, RI 02871

Corning Glass Works   52
EB-2 Houghton Park
Corning, NY 14830

Group Four Inc.   53
147 Simsbury Rd.
Avon, CT 06001

John Wistrand Design   54
74 Journey's End Rd.
New Canaan, CT 06840

Plumb Design Group   56
167 Third Ave.
New York, NY 10003

Texas Instruments   58, 59
P.O. Box 10508, MS 5819
Lubbock, TX 79408

JCPenney Company, Inc.   60
1301 Avenue of the Americas
New York, NY 10019

Tony Carsello Industrial Design   62
334 Figueroa St.
Wilmington, CA 90744

INOVA   63
530 Lytton Ave.
Palo Alto, CA 94301

Marcie Lipschutz   64
Philadelphia College of Art
1610 Naudain St.
Philadelphia, PA 19146

Keck-Craig Associates   66
245 Fair Oaks Ave.
South Pasadena, CA 91030

Contract and Residential

Deschamps ● Mills Assoc., Ltd.   73
864 Stearns Rd.
Bartlett, IL 60103

Ward Bennett   74
1 West 72nd St.
New York, NY 10023

Henry P. Glass Associates   76
P.O. Box 52
Northfield, IL 60093

Richard Schultz   77
Box 52
Barto, PA 19504

Interface Design Group, Inc.   78
205 West Highland Ave.
Milwaukee, WI 53203

George Pelling Design   79
1044 South Crescent Heights Blvd.
Los Angeles, CA 90035

Industrial Design Program   80
University of Notre Dame
P.O. Box 134
Notre Dame, IN 46556

Equipment and Instrumentation

Roger Lee/Design Associates   86
125 University Ave.
Palo Alto, CA 94301

Hewlett-Packard Company   88
3400 E. Harmony Rd.
Fort Collins, CO 80525

IBM Design Program   91, 92
Dept. 823, Bldg. 701
IBM Corporation
Poughkeepsie, NY 12602

Digital Equipment Corporation   93
Industrial Products Group
146 Main St.
Maynard, MA 01754

Bally Design, Inc.   94
219 Park Rd.
Carnegie, PA 15106

Goldsmith Yamasaki Specht Inc.   95, 99
840 North Michigan Ave.
Chicago, IL 60611

Carl Yurdin Industrial Design, Inc.   96
2 Harborview Rd.
Port Washington, NY 11050

Fisher Scientific Company   98
711 Forbes Ave.
Pittsburgh, PA 15219

Datascope Corp.   101
580 Winters Ave.
Paramus, NJ 07562

Rhode Island School of Design   103
Department of Industrial Design
2 College St.
Providence, RI 02903

Thermos Division of King-Seeley Thermos Co.   104
Norwich, CT 06360

Robert Hain Associates, Inc.   105
346 Park Ave.
Scotch Plains, NJ 07076

Richardson/Smith   108
Box 360
Worthington, OH 43085

Henry Dreyfuss Associates   110
888 Seventh Ave.
New York, NY 10019

Visual Communications

Group Four, Inc.   116
147 Simsbury Rd.
Avon, CT 06001

Copco, Inc.   117
11 East 26th St.
New York, NY 10010

Robert P. Gersin Associates, Inc.   118, 119, 137, 140
11 East 22nd St.
New York, NY 10010

Gerstman + Meyers Inc.   120, 121
46 West 55th St.
New York, NY 10019

Selame Design   122
2330 Washington St.
Newton Lower Falls, MA 02162

Design West, Inc.   123
2532 Dupont Dr.
Irvine, CA 92713

Morison S. Cousins + Associates Inc.   124
964 Third Ave.
New York, NY 10022

Anspach Grossman Portugal Inc.   125
850 Third Ave.
New York, NY 10022

Danne & Blackburn, Inc.   126, 136, 151-153
One Dag Hammarskjold Plaza
New York, NY 10017

JCPenney Company, Inc.   127–131
1301 Avenue of the Americas
New York, NY 10019

Container Corporation of America   132
Valley Forge Marketing Center
Oaks, PA 19456

Bill Bonnell   133, 134
7 West 14th St.
New York, NY 10011

Raychem Corporation   135, 142
Communications Design Center
300 Constitution Drive
Menlo Park, CA 94025

**Saul Bass/Herb Yager
and Associates**   138, 141, 148
7039 Sunset Boulevard
Los Angeles, CA 90028

ICOGRADA   144
7 Templeton Ct.
Radnor Walk
Croydon, England CRO 7 NZ

McCoy & McCoy   146, 147
500 Lone Pine Road
Bloomfield Hills, MI 48013

Gary Hanlon & Associates   149
1737 15th St.
Boulder, CO 80302

**Cook and Shanosky
Associates, Inc.**   150
221 Nassau St.
Princeton, NJ 08540

**Jonson Pedersen Hinrichs
& Shakery**   154
2 Embarcadero Center
San Francisco, CA 94111

Environmental

Norman Perman Design   163
233 East Erie St.
Chicago, IL 60611

**Smith, Hinchman
& Grylls Associates, Inc.**   164
455 West Fort St.
Detroit, MI 48226

George Nelson & Company   166
251 Park Avenue South
New York, NY 10010

**Robert P. Gersin
Associates, Inc.**   168-173, 177
11 East 22nd St.
New York, NY 10010

InterDesign Inc.   174
1409 Willow St.
Minneapolis, MN 55403

Walker/Group, Inc.   178
304 East 45th St.
New York, NY 10017

DIRECTORY OF PRODUCTS

Consumer Products

**Cuisinart Triomphe 500
Food Processor, Commercial** 50
Cuisinarts, Inc.
Stamford, CT 06712

**Corning Ware® MC-1 and
MC-2 Fast Food Dishes** 52
Corning Glass Works
Corning, NY 14830

**Thermos "Touch Top" Model 2647
Beverage Dispenser®** 53
King-Seeley Thermos Co.
Norwich, CT 06360

**Clairol Travel Hairdryer,
"One for the Road"** 54
Clairol
New York, NY 10022

Copal World Timer Alarm Clock 56
Copal Corporation of America
Woodside, NY 11377

**Texas Instruments Electronic
Analog Chronograph Watch,
Models 851 and 852** 58
Texas Instruments
Lubbock, TX 79408

**Texas Instruments "Speak & Spell"
Talking Learning Aid** 59
Texas Instruments
Lubbock, TX 79408

**JC Penney Hand-Held
Vacuum Cleaner** 60
JCPenney Company, Inc.
New York, NY 10019

**McCulloch Pro Mac 610
(Farmer) Chain Saw** 62
McCulloch Corporation
Los Angeles, CA 90066

**Prototype Aqua-Touch
Water Faucet Attachment** 63
Water-Plus
Los Altos, CA 94022

Prototype Hand-Held Hairdryer 64
Philadelphia College of Art
Philadelphia, PA 19146

Prototype Backpack with Cot 66
Jeffrey F. Jagels
Pasadena, CA 91101

Contract and Residential

Howell Deschamps I Chair 73
Howell Co.
St. Charles, IL 60174

Brickel 2412 Alexandria Chair 74
Brickel Associates, Inc.,
New York, NY 10022

**Brown Jordan "Cricket"
Lounge Chair** 76
Brown Jordan
El Monte, CA 91734

**Stow/Davis Paradigm Office
Seating** 77
Stow/Davis Furniture Co.
Grand Rapids, MI 49504

**Fiberesin Industries Fiberlife Line
of Eight Furniture Components** 78
Fiberesin Industries, Inc.
Oconomowoc, WI 53066

**Trakliting "Geometric"
Lighting Fixture** 79
Trakliting Inc.
City of Industry, CA 91746

Prototype Toilet 80
Industrial Design Program
University of Notre Dame
Notre Dame, IN 46556

Equipment and Instrumentation

**Hewlett-Packard 300
Computer System** 86
General Systems Division
Hewlett-Packard Co.
Santa Clara, CA 95050

**Hewlett-Packard 250
Business Computer System** 88
Hewlett-Packard Co.
Fort Collins, CO 80525

IBM 8775 Display Terminal 91
IBM Corporation
Armonk, NY 10504

**IBM 3616 Passbook
and Document Printer** 92
IBM Corporation
Armonk, NY 10504

**Digital Equipment RT803 and RT805
Factory Data Collection** 93
Terminal Workstations
Digital Equipment Corp.
Maynard, MA 01754

**MSA Lead-Foe 11 Air-Supplied
Hood** 94
Mine Safety Appliances Co.
Pittsburgh, PA 15208

**Encon 160 Chemical Splash
Goggle** 95
Encon (A Vallen Company)
Houston, TX 77001

Niranium Rollavue Sand Blaster 96
Niranium Corp.
Long Island City, NY 11106

**Fisher Accumet pH/Ion Meter,
Model 750** 98
Fisher Scientific Co.
Pittsburgh, PA 15219

GEX Intra-Oral Dental X-Ray 99
Medical Systems Division
General Electric Co.
Milwaukee, WI 53201

**Datascope M/D3 Defibrillator
System** 100
Datascope Corp.
Paramus, NJ 07652

**American Red Cross Mobile
Blood Collection System** 102
American Red Cross Blood Services
Washington, DC 20006

**Thermos Insulated Vaccine Carrier,
Model 3500/8706** 104
King-Seeley Thermos Co.
Norwich, CT 06360

**Medcor Lithicron-F
Pace Programmer** 105
Medcor, Inc.
Hollywood, FL 33023

**Picker Synerview 600
Computed Tomography
System** 106
Picker Corporation
Cleveland, OH 44143

**John Deere 2600 and 2800
Semi-Integral Moldboard Plow** 110
John Deere Plow/Planter Works
Moline, IL 61265

Visual Communications

**Dansk Teakwood Products
Packaging** 116
Dansk International Design
Mt. Kisco, NY 10549

**Copco Cutting Knives Sets
Packaging** 117
Copco, Inc.
New York, NY 10010

**Myojo Foods "O My Goodness"
Instant Oriental Noodle
Packages** 118
Myojo Foods of America, Inc.
New York, NY 10017

**Myojo Foods "O My Goodness"
Instant Oriental Noodle Marking
Dispenser Cartons** 119
Myojo Foods of America, Inc.
New York, NY 10017

**Pillsbury Panshakes
Pancake Mix Packages** 120
The Pillsbury Company
Minneapolis, MN 55402

**Royal Crown Cola's KICK
Packaging** 121
Royal Crown Cola Company
Schaumburg, IL 60195

**J.A Wright 7-lb Silver Polish
Can Graphics** 122
J.A. Wright & Company
Keene, NH 03431

**Rhodes "Beaver" and "Sunray"
Steel Wood Packages** 123
James H. Rhodes Co.
Des Plaines, IL 60018

**Gillette Max 1000
Hairdryer Package** 124
The Gillette Company
Boston, MA 02199

**McGraw-Edison Products
Packaging System** 125
McGraw-Edison Company
Elgin, IL 60120

**GAF Corporation GAFMED Brand
Identification and Packaging** 126
GAF Corporation
New York, NY 10020

**JCPenney's Automotive
Battery Line Graphics** 127
JCPenney Company, Inc.
New York, NY 10019

**JCPenney's Packaging
for Assorted Kitchen Items** 128
JCPenney Company, Inc.
New York, NY 10019

**JCPenney's Leotard
and Tights Packaging** 129
JCPenney Company, Inc.
New York, NY 10019

**JCPenney's Flasher Tags
for Boys' Shorts** 130
JCPenney Company, Inc.
New York, NY 10019

**JCPenney's Merchandise
Hang Tags for Children's
Coordinated Clothing** 131
JCPenney Company, Inc.
New York, NY 10019

**Container Corporation's
Shape Op® Container** 132
Carton Division
Container Corporation of America
Oaks, PA 19456

**Container Corporation's
Plastic Tank Sales Literature** 133
Plastics Division
Container Corporation of America
Chicago, IL 60670

**Container Corporation's 1978
Promotional Calendar** 134
Container Corporation of America
Chicago, IL 60670

**Raychem Corporate Identity
Guide** 135
Raychem Corporation
Menlo, Park, CA 94025

**IBM Corporate
Recruitment Literature** 136
IBM Corporation
Armonk, NY 10504

**Exhibition Catalog for the AIGA
"What's Real in Packaging"
Show** 137
American Institute of Graphic Arts
New York, NY 10021

**Girl Scouts Identification
Program** 138
Girl Scouts of the United States of
America
New York, NY 10022

**National Shirt Shops
Corporate Identity Program** 140
National Shirt Shops
New York, NY 10001

CBS: On the Air Trademark 141
CBS
New York, NY 10019

**Raychem Corporation Poster
Series** 142
Raychem Corporation
Menlo Park, CA 94025

**Three ICOGRADA Case Study
Posters** 144
ICOGRADA Chicago Congress
Croydon, England CRO 7 NZ

**Cranbrook Graduate Design
Poster** 146
Design Department
Cranbrook Academy of Art
Bloomfield Hills, MI 48013

**Cranbrook New York Trip
Poster** 147
Design Department
Cranbrook Academy of Art
Bloomfield Hills, MI 48013

"Bass on Titles" Poster 148
Pyramid Films
Santa Monica, CA 90406

Halleluiah Poster 149
Gary Hanlon & Associates
Boulder, CO 80302

**Cluett, Peabody & Co., Inc.
1977 Annual Report** 150
Cluett, Peabody & Co., Inc.
New York, NY 10036

**First Boston 1977 Annual
Report** 151
First Boston Corporation
New York, NY 10005

Potlach 1977 Annual Report 152
Potlach Corporation
San Francisco, CA 94119

**Hammermill Paper Company
1977 Annual Report** 153
Hammermill Paper Company
Erie, PA 16533

**Filmways Inc. 1978 Annual
Report** 154
Filmways, Inc.
Los Angeles, CA 90067

Environment

**American Hospital Supply
Corporation's Permanent
Lobby Exhibit** 160
American Hospital Supply
Corporation
Evanston, IL 60201

**Smith, Hinchman & Grylls
Foyer Light Display**   **164**
Smith, Hinchman & Grylls
Associates, Inc.
Detroit, MI 48226

**Banners for the
Aid Association for Lutherans**   **166**
Aid Association for Lutherans
Appleton, WI 54911

**Saint Francis Hospital
Exterior Signs**   **168**
Saint Francis Hospital
Tulsa, OK 74136

**Saint Francis Hospital
Main Directory**   **169**
Saint Francis Hospital
Tulsa, OK 74136

**Saint Francis Hospital
Interior Signs**   **170**
Saint Francis Hospital
Tulsa, OK 74136

**Saint Francis Hospital
Cancer Center Wall Graphic**   **172**
Saint Francis Hospital
Tulsa, OK 74136

**Minnesota Zoological Garden
Graphic Display System**   **174**
Minnesota Zoological Board
Apple Valley, MI 55124

National Shirt Shops' Interiors   **176**
National Shirt Shops
New York, NY 10001

I Natural Cosmetics Shop   **178**
I Natural Cosmetics
New York, NY 10022

Index

Adachi, Takashi, 118–119
Adasko, Laura, 166
Adou, Marsha C., 129
Anderson, Cynthia, 163
Anderson, David C., 63
Anselevicius, Evelyn, 166
Arai, Kiyoyuki, 56
Arnold, L.G., 110
Atseff, Laurence, 121

Bally, Alexander, 94
Banks, Seth, 99
Barbera, Larry, 108
Barnes, Jim, 99
Bass, Saul, 138, 141, 148
Bates, David, 163
Beckman, Frederick S., 80
Behnke, Bernice I., 103
Bell, Carolyn, 166
Bell, Richard L., 132
Bendon, Tom, 88
Bennett, Ward, 74
Benoit, Judy, 163
Best, Bob, 133
Bierman, Kerry, 163
Blackburn, Bruce, 126, 136
Blandine, Sister Mary, 168–173
Bleck, James A., 93
Bonnell, Bill, 131, 133–134
Brantingham, L., 59
Brunnett, Carl, 108
Buban, Elmer, 94
Burke, Joel E., 120

Campbell, Thomas J., 66
Carlson, Edward C., 53
Carsello, Tony, 62
Carter, Don F., 66
Caruso, Ingrid G., 168–173
Chamberlain, Edward, 90
Chang, R., 59
Clark, Casey, 118, 140
Colbert, Dave, 60
Colby, Donald B., 76
Concepcion, Juan, 121
Connolly, Peter F., 105
Cook, Roger, 150
Cooke, Kenneth R., 137, 140, 168, 171
Crettol, Gabrielle, 169, 171, 177
Cunningham, Tim, 94
Cusack, Margaret, 166

Dahl, Gene, 58
Daley, Barbara, 171, 173
Danne, Richard, 151–153
Dantzler, Justin L., 63
Degen, Robert W., 168–173
Deschamps, Robert Louis, 73
Drobeck, Bob, 64
Drozdowski, Robert J., 105

Eaton, David W., 116
Edwards, Larry W., 52
Erickson, David L., 78

Fattal, Vahe, 138
Febvre, Paul, 88
Feliciano, Rafael, 121
Finesman, Al, 50
Frantz, G., 59
Friberg, Kristina, 166
Friedman, Leonard I., 103
Fujitaki, Roy K., 66
Fuller, George W., 53
Fusco, Paul, 151
Fyfield, R.W., 54

Gaber, Ira, 127
Gagne, Roger, 93
Gall, John, 73
Garrecht, Alea, 140
Gersin, Robert P., 118–119, 137, 140, 168–173, 177
Gerstman, Richard, 120
Gibboney, Dr. Dennis, 98
Glass, Henry P., 76
Gluck, Nathan, 137
Goldschmidt, James, 168, 171
Goodman, Art, 138
Goodson, Marlowe, 151
Graham, Donald R., 98
Greiner, Ania, 144
Greiner, John, 144
Grossman, Eugene J., 125
Grube, Gary, 163
Grubel, Bob, 58
Gulick, Ken, 93

Hacker, Chris, 60
Hacker, J. Christopher, 116
Hadfield, Peter J., 104
Hadtke, Frederick B., 105
Hain, Robert W., 105
Hanlon, Gary, 149
Harrison, Marc S., 50, 103
Hawkins, W.R., 59
Hayashi, Katsuhiko, 118–119
Kinrichs, Kit, 154
Hobor, Nancy, 163
Hoffman, Jenet, 166
Holmquist, Anders, 166
Hopkins, Ken, 163
Horine, David, 86
Hottes, Ronald W., 62
Hurtado, Rich, 123

Jackson, W.W., 110
Johnson, Carl L., 53

Katz, Marjorie, 128–130
Kaulfuss, Bill, 134

Keck, Henry C., 66
Khovaylo, Mo, 119
Kipp, Michael A., 132
Kirsch, Jeanne, 60
Kitts, Leonard W., 78
Klein, Richard, 135
Kobold, Elaine, 163
Koester, Frank, 180
Kortlander, Ann, 135
Koslow, Howard, 166
Kovner, Jules, 140
Kremer, William, 80
Kresge, Keith, 108
Kuyt, Frits, 58

LaLiberte, Norman, 166
Lane, Judith, 177
Laude, Michael E., 104
Law, David, 129–130
Lawing, Edward, 103, 108
Lawrence, Karen, 166
Lawrence, W.J., 58–59
Leaf, Georgina, 118, 169, 171
Lee, Roger, 86
Lewin, Gideon, 129
Lipschutz, Marcie, 64
Long, Nancy, 135
Lotito, Frank, 94
Luttmer, Joan, 80
Lyons, Richard, 73

Macconkey, James S., 93
McCoy, Katherine, 146–147
McCoy, Michael, 146–147
McDonagh, Jeanne, 166
McGhie, Bruce, 150
Maffey, George E., 62
Mann, Douglass, 96
Marcotte, John, 58
Markowitz, Daniel, 140, 177
Marks, Burt, 58
Marks, Edward, 140, 177
Marks, Steven, 140, 177
Mathis, Barry, 88
Meiser, Ken, 60
Meyers, Herbert M., 121
Miller, Paul, 126
Muller, Lois, 180

Nakano, Kyoji, 163
Neil, George A., 52
Nelson, Louis, 118–119, 140, 177
Nesbitt, R., 59
Nishiyama, Takashi, 118–119
Noble, Robert, 58

Oelschlaeger, George, 110
Onihsi, S., 56

Palazzolo, Frank, 60

Pandya, Roxane, 129
Passy, Marc-Albert, 117
Pelling, George E., 79
Perman, Norman, 163
Phillips, Bill, 94
Piegdon, Ted, 123
Pierce, Ron, 60, 127
Purcell, William F.H., 110
Purdy, Robert, 98

Raina, Ken, 93
Ratkoi, George, 116
Redding, Robert M., 132
Rein, Hal, 163
Rich, Jeff, 163
Richardson, Deane, 108
Riddell, Larry, 120
Rieben, John R., 135, 142
Robert, Francois, 144
Rogalski, Jeff, 125
Rongren, Bev, 175
Ross, Jerrold, 116
Rossum, Cheryl, 153
Roswog, Ed, 99

Salisbury, Tom, 163
Savage, Carol, 118–119
Scharples, Sue, 105
Schlesinger, David, 101
Schneider, John, 98
Schnipper, Steven, 130
Schory, Kenneth, 73
Schultz, Richard, 77
Schumacher, Johann, 124

Schuman, Peter, 105
Schwartz, Dina, 166
Schwartz, Robert T., 103
Seager, Richard H., 104
Seitz, Peter, 175
Serbinski, Andrew T., 56
Selame, Joseph, 122
Selame, Robert, 122
Sferva, Richard, 146
Sharp, Mike, 90
Shea, Owen, 92
Shoulberg, Sam, 125
Silenski, Ed, 120
Smith, Phil, 163
Sontheimer, Carl, 50
Specht, Paul B., 95, 99
Stanley, Richard, 135
Shanosky, Don, 150
Shapiro, Barbara, 154
Stannard, Steve, 164
Stein, John, 163
Stein, Sanford, 175
Sterling, David, 147
Sterling, Joseph, 163
Stickney, Joe, 108
Stillinger, Scott, 86
Stone, David, 166
Stone, Robert, 177
Stout, Lee, 140, 177
Suiter, Tom, 123
Sulek, E.J., 58–59

Tarozzi, Richard A., 53, 104
Terrell, Douglas H., 116

Terrell, William, 101
Tompkins, Dave, 108
Tompkins, Dr. Robert G., 168–173
Tracy, Tom, 154
Tribbey, Jan, 60
Turner, Robert L., 180

Uzzle, Burk, 152

Verret, Lloyd J., 168–173
Virgilio, Pam, 119
Vogele, Robert, 144

Walker, Kenneth H., 180
Weiner, Jack, 154
Whitney, Patrick, 144
Wilson, Tim, 60
Winer, Mel, 163

Wistrand, John, 54
Witt, Marilee, 129
Wong, Ronald, 137
Wright, Robert, 95

Yabu, Kiyohisa, 56
Yamamoto, Hideki, 175
Yoshi, 166
Younge, P. Dennis, 52
Yurdin, Carl, 96

Zimmerer, John L., 62
Zimmerman, Carol, 128
Zupanick, Tony, 108

Edited by Sarah Bodine and Susan Davis
Designed by Jay Anning
Set in 10 point Palatino